Naming Rights, Place Branding, and the Cultural Landscapes of Neoliberal Urbanism

In recent decades, urban policymakers have increasingly embraced the selling of naming rights as a means of generating revenue to construct and maintain urban infrastructure. The contemporary practice of toponymic commodification has its roots in the history of philanthropic gifting and the commercialization of professional sports, yet it has now become an integral part of the policy toolkit of neoliberal urbanism more generally. As a result, the naming of everything from sports arenas to public transit stations has come to be viewed as a sponsorship opportunity, yet such naming rights initiatives have not gone uncontested.

This edited collection examines the political economy and cultural politics of urban place naming and considers how the commodification of naming rights is transforming the cultural landscapes of contemporary cities. Drawing upon case studies ranging from the selling of naming rights for sports arenas in European cities and metro stations in Dubai to the role of philanthropic naming in the "Facebookification" of San Francisco's gentrifying neighborhoods, the contributions to this book draw attention to the diverse ways in which toponymic commodification is reshaping the identities of public places into time-limited, rent-generating commodities and the broader implications of these changes on the production of urban space.

The chapters in this book were originally published as a special issue of *Urban Geography*.

Reuben Rose-Redwood is Professor in the Department of Geography at the University of Victoria in British Columbia, Canada.

Jani Vuolteenaho is Senior Lecturer in the Faculty of Social Sciences at the University of Helsinki in Finland.

Craig Young is Professor in the Department of Natural Sciences at Manchester Metropolitan University in the United Kingdom.

Duncan Light is Senior Lecturer in Tourism at Bournemouth University in the United Kingdom.

Naming Rights, Place Branding, and the Cultural Landscapes of Neoliberal Urbanism

Edited by
Reuben Rose-Redwood, Jani Vuolteenaho, Craig Young, and Duncan Light

Routledge
Taylor & Francis Group
LONDON AND NEW YORK

First published 2022
by Routledge
2 Park Square, Milton Park, Abingdon, Oxon OX14 4RN

and by Routledge
605 Third Avenue, New York, NY 10158

Routledge is an imprint of the Taylor & Francis Group, an informa business

British Library Cataloguing in Publication Data
A catalogue record for this book is available from the British Library

ISBN: 978-0-367-75624-6 (hbk)
ISBN: 978-0-367-75626-0 (pbk)
ISBN: 978-1-003-16326-8 (ebk)

Typeset in Minion Pro
by Newgen Publishing UK

Publisher's Note
The publisher accepts responsibility for any inconsistencies that may have arisen during the conversion of this book from journal articles to book chapters, namely the inclusion of journal terminology.

Contents

Citation Information

The following chapters were originally published in *Urban Geography*, volume 40, issue 6 (2019). When citing this material, please use the original page numbering for each article, as follows:

Preface
Reuben Rose-Redwood, Jani Vuolteenaho, Craig Young and Duncan Light

Introduction
Naming rights, place branding, and the tumultuous cultural landscapes of neoliberal urbanism
Reuben Rose-Redwood, Jani Vuolteenaho, Craig Young and Duncan Light
Urban Geography, volume 40, issue 6 (2019), pp. 747–761

Chapter 1
"This venue is brought to you by. . .": the diffusion of sports and entertainment facility name sponsorship in urban Europe
Jani Vuolteenaho, Matthias Wolny and Guy Puzey
Urban Geography, volume 40, issue 6 (2019), pp. 762–783

Chapter 2
Scalar tensions in urban toponymic inscription: the corporate (re)naming of football stadia
Dominic Medway, Gary Warnaby, Leah Gillooly and Steve Millington
Urban Geography, volume 40, issue 6 (2019), pp. 784–804

Chapter 3
Who owns the name? Fandom, social inequalities and the contested renaming of a football club in Timişoara, Romania
Remus Creţan
Urban Geography, volume 40, issue 6 (2019), pp. 805–825

Chapter 4

Data, dispossession, and Facebook: techno-imperialism and toponymy in gentrifying San Francisco
Erin McElroy
Urban Geography, volume 40, issue 6 (2019), pp. 826–845

Chapter 5

"Turn your brand into a destination": toponymic commodification and the branding of place in Dubai and Winnipeg
Reuben Rose-Redwood, Maral Sotoudehnia and Eliot Tretter
Urban Geography, volume 40, issue 6 (2019), pp. 846–869

Chapter 6

City renaming as brand promotion: exploring neoliberal projects and community resistance in New Zealand
Robin A. Kearns and Nicolas Lewis
Urban Geography, volume 40, issue 6 (2019), pp. 870–887

Afterword

The names of urban dispossession: a concluding commentary
David J. Madden
Urban Geography, volume 40, issue 6 (2019), pp. 888–892

Notes on Contributors

Remus Creţan, Department of Geography, West University of Timişoara, Timişoara, Romania.

Leah Gillooly, Sport Policy Unit, Manchester Metropolitan University, Manchester, UK.

Robin A. Kearns, School of Environment, University of Auckland, Auckland, New Zealand.

Nicolas Lewis, School of Environment, University of Auckland, Auckland, New Zealand.

Duncan Light, Department of Tourism and Hospitality, Bournemouth University, Bournemouth, UK.

David J. Madden, Department of Sociology, London School of Economics and Political Science, London, UK.

Erin McElroy, Feminist Studies, University of California, Santa Cruz, CA, USA.

Dominic Medway, Institute of Place Management, Manchester Metropolitan University, Manchester, UK.

Steve Millington, Sport Policy Unit, Manchester Metropolitan University, Manchester, UK.

Guy Puzey, Scandinavian Studies, School of Literatures, Languages and Cultures, University of Edinburgh, Edinburgh, Scotland.

Reuben Rose-Redwood, Department of Geography, University of Victoria, Victoria, Canada.

Maral Sotoudehnia, Department of Geography, University of Victoria, Victoria, Canada.

Eliot Tretter, Department of Geography, University of Calgary, Calgary, Canada.

Jani Vuolteenaho, Faculty of Social Sciences, University of Helsinki, Helsinki, Finland.

Gary Warnaby, Institute of Place Management, Manchester Metropolitan University, Manchester, UK.

Matthias Wolny, Department of Romance Studies, Faculty of Modern Languages, Heidelberg University, Heidelberg, Germany.

Craig Young, Division of Geography and Environmental Management, Manchester Metropolitan University, Manchester, UK.

Preface

Black Lives Matter Plaza. George Perry Floyd Jr Place. Denali National Park. Mohamed Bouazizi Square. The naming of places – whether they be streets, squares, parks, or other locales – matters. Place names shape our geographical imaginations and not only help us navigate within our physical surroundings but also provide geographical anchor points to ground our social and political values in the spaces of everyday life. Those in positions of political authority have long understood the power of place naming and have sought to rename the world in their own image. At the same time, groups who have been socially marginalized and politically oppressed know all-too-well how place naming also produces spaces of exclusion and erasure.

There has been a growing movement in recent years to contest the exclusions of white supremacist and masculinist commemorative practices by renaming places as well as removing statues and monuments honoring white supremacists, slave traders, and colonizers while calling for the commemoration of racialized minorities, indigenous peoples, and women. In Bristol, UK, for example, the statue of seventeenth-century merchant and slave trader Edward Colston was torn down and dumped into the harbor by Black Lives Matter demonstrators in June 2020 and was temporarily replaced by a statue of contemporary protester Jen Reid, a woman of color who had been photographed standing on the empty plinth with her fist raised (Bland, 2020). In the United States, pressure has grown to change the names of US military bases which bear the names of Confederate generals from the US Civil War (1861–65) – Forts Bragg, Lee, Pickett, Hood, Polk, and Rucker. Together with statues, monuments, and memorials, place naming has therefore played an important role in the making, unmaking, and remaking of commemorative landscapes, which are generally produced to legitimize particular sets of political ideologies and economic practices.

Changes to the cultural landscape often arise during times of political upheaval, regime change, or revolution, yet they also occur as part of the ordinary workings of municipal government and grassroots organizing at the local scale. The act of commemorative naming is *honorific* – that is, it honors a person, place, event, or concept that is deemed worthy of public recognition. Commemorative place naming has commonly been used to honor those viewed as the "heroes" of the city, nation, or empire. Following the terrorist attacks of September 11, 2001, for instance, hundreds of streets in New York City were given honorary names recognizing first responders who died in the attacks (Sharp, 2014). More recently, in the midst of the current COVID-19 pandemic, memorials and monuments to the victims of the global pandemic are already in the works (Associated Press, 2020;

Baldwin, 2020), with one 20-foot statue in Riga, Latvia, of a female medical worker in personal protective equipment (PPE) named "Medics to the World" erected in honor of health professionals (Tidman, 2020). In a similar manner, but with a decidedly economic turn, in Moscow e-commerce and food-tech companies paid for a new statue honoring delivery workers who made self-isolation possible during lockdown, interestingly located at a business center (Kishkovsky, 2020). It is just a matter of time before streets and other places in cities around the world are renamed to honor front-line workers and others who have died as a result of the coronavirus pandemic as well.

Yet while one of the primary uses of place naming has been to inscribe the values of a society or community into the cultural landscape through honorific commemoration, over the past few decades the *economic* function of place naming has taken on increasing precedence. In October 2020, the inhabitants of the small mining town of Asbestos in Quebec, Canada, voted to rename the town Val-des-Sources to shed its association with what had been the world's largest asbestos mine and the cancer-causing mineral. Though this renaming involved an economic cost in itself (estimated at US$100,000), it was felt that the negative perceptions aroused by the town's name and its industrial history were deterring inward investment, and the renaming was vital for its future economic prosperity (BBC, 2020; Cox, 2020). As this example shows, naming can play an integral role in the process of *place branding* as an economic development strategy.

Furthermore, going beyond renaming places as part of efforts to sustain economic development, recent decades have also witnessed a growing number of public authorities selling the legal naming rights for everything from sports arenas to metro stations to the highest bidder. This process of "toponymic commodification" has proliferated globally as part of the turn toward neoliberal modes of urban governance where the identities of public places are recast by urban policymakers as potential rent-generating assets rather than public goods. In many contexts, global economic crises have led to the imposition of austerity politics and budgetary restrictions, forcing urban authorities to look for new sources of revenue. The commodification of naming rights is believed by some to offer one potential solution, but it also has various other implications within the neoliberalized city, such as the privatization of public space and the opportunity costs associated with foreclosing the prospects of using place naming to advance movements for social and spatial justice.

The selling of public naming rights to private actors reflects the unequal distribution of political and economic power and, in turn, potentially further entrenches existing power relations within capitalist societies. The naming of public places by particular corporate interests or wealthy individuals has the effect of normalizing the unequal and potentially exploitative economic relations that underpin their wealth in the everyday landscape, thereby obscuring the power, inequalities, and environmental harm at play. And this is happening at the same time as contemporary social movements seek to decolonize commemorative landscapes and expose the economic relationships (such as slavery or exploitative forms of capitalism) that are implicit within them.

To date, few urban scholars have critically analyzed the political economy and cultural politics of commodifying naming rights. One of the chief aims of this edited book is therefore to showcase scholarship that brings a critical perspective to bear on the commodification of urban naming rights and its implications for the social production of urban cultural spaces. As such, this collection contributes not only to critical studies of place naming but also seeks to bring together the insights of political economy and cultural politics to advance critical urban studies more generally.

In the current era of global crises, the rise of various forms of authoritarian neoliberalism – where neoliberal reforms and authoritarian statisms intertwine (see Tansel, 2017; Bruff and Tansel, 2019) – has led to an intensification of so-called "culture wars" that are creating a variety of new contexts in which cultural practices (including naming) and political economy intersect in complex ways (e.g. Grcheva, 2019). This fast-changing world calls for fresh critical-theoretical approaches and contextually sensitive perspectives to grasp these transformations, offering emerging avenues for advancing critical toponymic theory and praxis. In the pages that follow, the contributors to the present book navigate through these complex entanglements of culture, economy, and politics as they coalesce and collide in the processes of toponymic commemoration and commodification.

Reuben Rose-Redwood, Jani Vuolteenaho, Craig Young and Duncan Light
November 2020

References

Associated Press. (2020). "Brazil: Cemetery Monument Commemorates COVID-19 Victims." *The Star*, September 20. Accessible at:www.thestar.com/news/world/americas/2020/09/20/brazil-cemetery-monument-commemorates-covid-19-victims.html (accessed on November 16, 2020).

Baldwin, Eric. (2020). "World's First Large-Scale COVID Memorial Designed for Victims of the Pandemic." *Arch Daily*, August 19. Accessible at: www.archdaily.com/945873/worlds-first-large-scale-covid-memorial-designed-for-victims-of-the-pandemic (accessed on November 16, 2020).

BBC. (2020). "New Name for a Canadian Town called Asbestos." *BBC News*, October 19. Accessible at: www.bbc.co.uk/news/world-us-canada-54608836 (accessed on November 16, 2020).

Bland, Archie. (2020). "Edward Colston Statue Replaced by Sculpture of Black Lives Matter Protester Jen Reid." *The Guardian*, July 15. Accessible at:www.theguardian.com/world/2020/jul/15/edward-colston-statue-replaced-by-sculpture-of-black-lives-matter-protester (accessed on November 16, 2020).

Bruff, Ian and Tansel, Cemal Burak. (2019). "Authoritarian Neoliberalism: Trajectories of Knowledge Production and Praxis." *Globalizations* 16(3): 233–244.

Cox, David. (2020). "Why the Canadian Town of Asbestos Wants a New Name." *BBC Global News*, October 13. Accessible at: www.bbc.com/worklife/article/20201008-why-the-canadian-town-of-asbestos-wants-a-new-name (accessed on November 16, 2020).

Grcheva Leonora. (2019). "The Birth of a Nationalistic Planning Doctrine: The 'Skopje 2014' Project." *International Planning Studies* 24(2): 140–155.

Kishkovsky, Sophia. (2020). "Signed, Sealed, Delivered: New Monument Honours Couriers on the Front Line of the Coronavirus Pandemic." *The Art Newspaper*, July 3. Accessible at:www.theartnewspaper.com/news/courier-monument (accessed on November 16, 2020).

Sharp, Sonja. (2014). "An Interactive Map of All the NYC Streets Named After 9/11 Victims." *Vice*, September 11. Accessible at: www.vice.com/en/article/mv5xbv/the-ghost-streets-of-911-009 (accessed on November 16, 2020).

Tansel, Cemal Burak (ed.). (2017). *States of Discipline: Authoritarian Neoliberalism and the Contested Reproduction of Capitalist Order*. London: Rowman & Littlefield International.

Tidman, Zoe. (2020). "Latvia Unveils Large Statue Honouring Covid-19 Healthcare Workers Around the World." *The Independent*, June 19. Accessible at:www.independent.co.uk/news/world/europe/coronavirus-latvia-statue-healthcare-workers-honour-a9574836.html (accessed on November 16, 2020).

Introduction

Naming rights, place branding, and the tumultuous cultural landscapes of neoliberal urbanism

Reuben Rose-Redwood, Jani Vuolteenaho, Craig Young and Duncan Light

ABSTRACT

In recent decades, urban policymakers have increasingly embraced the selling of naming rights as a means of generating revenue to construct and maintain urban infrastructure. This practice of "toponymic commodification" first emerged with the commercialization of professional sports during the second half of the 20th century and has become an integral part of the policy toolkit of neoliberal urbanism more generally. As a result, the naming of everything from sports arenas to public transit stations has come to be viewed as a sponsorship opportunity, yet such naming rights initiatives have not gone uncontested. This special issue examines the political economy of urban place naming through a series of case studies that consider how the commodification of naming rights is transforming the cultural landscapes of contemporary cities. In this introductory article, we provide an overview of the geographies of toponymic commodification as an emerging research focus within the field of critical urban toponymies and propose several theoretical approaches that can enhance our understanding of the commodification of naming rights as an urban spatial practice. We then discuss the main contributions in this special issue and conclude by exploring potential directions for future research on the geographies of urban toponymic commodification.

Introduction

On 1 December 2016, the Los Angeles County Metropolitan Transportation Authority (MTA) approved a policy allowing corporate sponsors to purchase the right to rename "a rail line, bus line, station or building by stamping their name and logo all over the facility, across station signs and locator graphics as well as digital incarnations such as Google maps or other online directional services" (Scauzillo, 2016). Within less than three months, the very same governing body repealed its new naming rights policy due to concerns that it would result in the over-commercialization of public space and potentially lead to lawsuits if the MTA denied a corporation's naming rights proposal (Nelson, 2017; Scauzillo, 2017). Meanwhile, in Saudi Arabia's capital city, Riyadh, governmental authorities announced in 2018 that they had acquired over $200 million (USD) in naming rights contracts for eight metro stations (Bridge, 2018), thereby joining a growing number of cities worldwide – from New York to Dubai – that have embraced naming rights sponsorships as a means of generating revenue for the development and maintenance of public infrastructure.

Historically, the selling of naming rights for urban landmarks traces back to two primary sources: *philanthropic gifting* and the *commercialization of professional sports*. Over the past two decades, however, the commodification of naming rights has become a seemingly ubiquitous practice with city governments, transportation authorities, hospitals, universities, environmental conservation groups, and other non-governmental organizations selling the naming rights for everything from convention centers and sports arenas to public parks and even newly discovered species (Burton, 2008). As a form of public-private partnership, naming rights sponsorships are a political technology of neoliberal governance that have played a significant role in the commodification of urban place-identities. The process of *toponymic commodification* – that is, the use of place names, or toponyms, as commodities – is radically transforming the identities of public places into marketized assets, the value of which is increasingly being framed solely in terms of a name's rent-generating capacity.

Proponents argue that the sale of naming rights offers an innovative strategy to generate municipal revenue without raising taxes, whereas scholars and grassroots activists in different urban and national contexts have maintained that naming rights sponsorships commercialize, and thus erode the democratic value of, urban public spaces. Although there is a long history of places being named after wealthy elites, property owners, and corporations, the contributions to this special issue illustrate that the proliferation of naming rights agreements in recent decades is historically unprecedented in terms of its scope and the interest in toponymic commodification within urban policymaking circles. Naming rights contracts have also changed as "time-limited" naming rights agreements have become the norm in comparison with more traditional forms of philanthropic naming in which an honorific name is bestowed in perpetuity. The time-limited nature of most contemporary naming rights agreements has resulted in urban namescapes that remain in a constant state of flux as one corporate sponsor's name replaces another when naming rights contracts expire. These tumultuous changes in the place-identities of major urban landmarks have significant implications for urban wayfinding and the creation of a "sense of place" in contemporary cities.

Despite its growing importance as an urban policy practice, few urban scholars have critically examined the geographies of urban naming rights and toponymic commodification (yet, see Light & Young, 2015; Medway & Warnaby, 2014; Rose-Redwood, 2011; Rose-Redwood, Alderman, & Azaryahu, 2010). The primary goal of this special issue is to showcase empirically-grounded and theoretically-informed case studies of how the commodification of naming rights and the use of place naming as a city branding strategy have become integral to the workings of neoliberal urbanism. Our overarching aim is therefore to contribute to geographical scholarship on urban neoliberalization by examining how the processes of commodification have extended into controversial new arenas of urban place-making that are reshaping the cultural landscapes of cities around the world.

In the remainder of this introductory article, we situate recent scholarship on toponymic commodification within the broader literature on critical toponymies.

Next, we consider various theoretical approaches that have been, or might productively be, employed to study the geographies of toponymic commodification. We then discuss the main contributions that this special issue makes toward developing a body of urban geographical scholarship on the commodification of naming rights and its role in the neoliberalization of urban space. Lastly, the article concludes by suggesting potential directions for future research on urban toponymic commodification as a spatial strategy of neoliberal urbanism.

Toponymic commodification and the geographies of urban naming rights: an emerging field of study

The study of toponymic commodification is an emerging area of scholarship within the field of critical urban toponymy. The geography of place naming has long been of interest to geographers and other scholars, and there is a well-established body of literature which has explored the important role that naming plays in the production of space and place. Early toponymic scholarship primarily adopted encyclopedic, etymological, and archaeological approaches, resulting in gazetteers and dictionaries of place names as well as using toponymy to reconstruct the history of settlement within a region and the influence of past cultures on the landscape (for a review of early place name studies, see Wright, 1929). In recent years, however, the study of place naming has been reanimated through the emergence of *critical toponymies*, which refers to a growing body of research that critically examines the politics of place naming practices (Berg & Vuolteenaho, 2009; Giraut & Houssay-Holzschuch, 2016; Rose-Redwood et al., 2010).

During the 1990s, a number of seminal works highlighted how the naming of places, streets, and urban landmarks was implicated in political projects such as nation-building, state formation, and the spatialization of collective memory (Azaryahu, 1996; Berg & Kearns, 1996; Cohen & Kliot, 1992; Myers, 1996; Yeoh, 1996). Building on these classic studies, a new wave of critical toponymic scholarship extended this work by analyzing a range of case studies related to the politics of (re)naming places in a variety of geographical contexts, with a particular focus on colonial/postcolonial, post-Apartheid, and socialist/postsocialist settings (Bigon, 2009; Duminy, 2014; Light, 2004; Light & Young, 2014; Wanjiru & Matsubara, 2017). A parallel body of research has also linked naming and renaming to the politics of race, gender, class, and the geographies of social justice (Alderman, 2002; Alderman & Inwood, 2013; Rose-Redwood, 2008; Rose-Redwood et al., 2010; Rose-Redwood, Alderman, & Azaryahu, 2018).

Over the past decade, there has been increasing attention not only to the politics of place naming but, more specifically, to the *political economy* of toponymic commodification (Light & Young, 2015; Medway & Warnaby, 2014; Rose-Redwood & Alderman, 2011; Rose-Redwood et al., 2010; Sotoudehnia & Rose-Redwood, 2019). This focus on the "toponym-as-commodity" (Rose-Redwood & Alderman, 2011, p. 3) conceptualizes place naming as a form of symbolic capital, which can be converted into economic capital and exploited for its exchange value (Alderman, 2008; Light & Young, 2015; Rose-Redwood et al., 2010). In particular, Rose-Redwood (2011) argues that the commercialization of urban place naming within contemporary neoliberal contexts is one of the most significant transformations that will affect the toponymic landscape during the 21st century (also, see Light & Young, 2015).

This focus on the commodification of urban toponymy thus constitutes an embryonic "second wave" of critical place name scholarship to which the current special issue contributes. The selling of naming rights for sports stadia to corporate sponsors – itself one of the longest-established ways in which place names have been turned into commodities – has been examined from a range of disciplinary perspectives including sports management (e.g. Eddy, 2014) and marketing (e.g. Woisetschläger, Haselhoff, & Backhaus, 2014). Yet only recently have critical place name scholars begun to consider the commodification of sports stadia naming, the motives of the clubs themselves, and the reactions of ordinary fans (Gillooly & Medway, 2019; Light & Young, 2015; Medway & Warnaby, 2014; Vuolteenaho & Kolamo, 2012). Several of the articles in the current special issue further advance this line of scholarly inquiry.

Increasingly, geographers have begun to consider how toponymic commodification has extended beyond the realm of professional sports to become a political technology of neoliberal urbanism more generally (Light & Young, 2015; Medway & Warnaby, 2014; Sotoudehnia & Rose-Redwood, 2019). Such studies examine how municipal authorities have embraced the selling of naming rights to public places – such as the naming of metro stations, public buildings, and other elements of a city's public infrastructure – as a means of coping with declining revenue transfers from central governments, austerity politics, global financial crises, and a prevailing neoliberal orthodoxy emphasizing the need for cities to be entrepreneurial and competitive (Light & Young, 2015; Rose-Redwood, 2011). A related strand of research considers the influence of neoliberal urbanism on the (re) naming of urban neighborhoods, particularly in the context of private sector-led urban development projects (Berg, 2011; Madden, 2018; Masuda & Bookman, 2018; Vuolteenaho & Ainiala, 2009), thereby emphasizing struggles over access to, and belonging in, urban space in the neoliberal city.

Some studies have also focused on toponymic commodification as a form of city branding (Medway & Warnaby, 2014; Sotoudehnia, 2013; Sotoudehnia & Rose-Redwood, 2019). Medway and Warnaby (2014), for example, argue that toponyms can serve as the basis for place brands but cannot be treated in the same way as conventional brand names. Moreover, place brands can be dissonant with both internal and external perceptions of a place (Medway, Swanson, Neirotti, Pasquinelli, & Zenker, 2015). This points to the need for a deeper understanding of how place brands – and names – resonate with the stakeholders involved in both their creation and consumption. Addressing this latter issue, Hakala, Sjöblom, and Kantola (2015) explore the perceptions of local urban residents towards the renaming of municipalities in Finland. Their research suggests that local residents often show a strong attachment to existing place names and that stable names are essential in the construction of place brands.

The commodification of place naming also occurs in more idiosyncratic ways. For instance, Light (2014) examines the commodification of place naming in the context of tourism geographies and argues that unusual toponyms can themselves be attractions for tourists. Furthermore, place names are commodified in a diverse range of souvenirs (such as fridge magnets), which enable tourists to make connections with the places they visit. In a different context, Karimi (2016) considers place names as commodities in post-conflict Kabul but argues that such names were valued more as symbolic rather than economic capital. Finally, Sotoudehnia (2018) explores the adoption of "Makani" (a wayfinding tool based on geo-addressing software) in Dubai. She argues that this app represents a unique example of the "toponym-as-commodity" (Rose-Redwood & Alderman, 2011) in that the data collected by the app could become commodities for sale to third parties, thereby contributing to the monetization of urban space. Similarly, Light and Young (2015) make related arguments about the privatization and commercialization of spatial data that were originally produced by public sector organizations.

As this review of the literature on the political economy of urban place naming illustrates, the commodification of urban toponymy is a key emerging research area. To date, the focus of this second wave of critical toponymic research is eclectic and fragmented, and this special issue aims to bring together a series of theoretically-informed case studies within a single volume to further enrich our understanding of the geographies of toponymic commodification in neoliberalizing cities.

Theoretical perspectives on the political economies of urban place naming

In response to the global upsurge in the commodification of naming rights as part of the neoliberalization of urban space, it is crucial for urban scholars to consider what different theoretical frameworks can offer for describing, interpreting, and critiquing the political economies of urban place naming. In this section, we briefly consider how theories of capitalist urbanization, symbolic capital, hegemony, and rhythmanalysis can provide insights into the production of commodified toponymic landscapes. These approaches are by no means the only theoretical frameworks available to examine the political economies of urban place naming. This theoretical discussion is therefore meant to be suggestive rather than comprehensive, and we hope that it will inspire greater theoretical reflection in subsequent geographical studies of toponymic commodification.

One of the most relevant sources of theoretical inspiration to critically examine the geographies of commodified urban toponymies are historical-geographical materialist accounts of capitalist urbanization. As Harvey (1985) argues, urbanization under capitalism is based upon the imperative to "create a social and physical landscape in its own image . . . only just as certainly to undermine, disrupt and even destroy that landscape at a later point in time" (p. 150). From this perspective, the selling of naming rights to urban infrastructure can be understood as an extension of commodity relations to the realm of urban place-identities. In particular, toponymic commodification is a rent-seeking practice that depends upon the exercise of a monopoly power over the naming of places (for a discussion of the concept of monopoly rent, see Harvey, 2012, pp. 89–112).

Both philanthropic and corporate-sponsored naming rights involve the exercise of a monopoly power over property (whether public or private), but not all forms of property function as marketized commodities. Philanthropic naming has historically involved a process whereby a donor provides a gift to an organization in return for naming rights *in perpetuity*. While the initial competition for securing a philanthropic naming right can be seen as a form of toponymic commodification, once a philanthropic naming right is bestowed, this generally has the effect of removing it from the potential market of commodified naming rights indefinitely. By contrast, time-limited naming rights sponsorships ensure that when naming rights are leased to sponsors for a given period of time, they will eventually be released back onto the naming rights market when the lease expires. The time-limited quality of naming rights sponsorships transforms a naming right into a *rent-generating, marketable commodity*.

A significant geographical consequence of the shift to time-limited naming rights is the destabilization of urban namescapes, since corporate-sponsored place names often change when naming rights contracts are renegotiated, thereby contributing to what urban geographers have called the "restless" urban landscape (Knox, 1991). Toponymic commodification is thus part of the broader process of creative destruction that characterizes capitalist urbanization, which, as Harvey contends, "must negotiate a knife-edge between preserving the values of past commitments … or devaluing them to open up fresh room for accumulation" (1985, p. 150). The selling of time-limited naming rights devalues "past commitments" of place naming in order to "open up fresh room for accumulation" for both public and private property owners while also generating public visibility in the city for corporate brands as part of a sponsor's marketing strategy. It is hardly surprising, then, that the proliferation of commodified naming rights has gone hand-in-hand with the rise of "entrepreneurial municipalities" that have internalized neoliberal imperatives (Lauermann, 2016). Historical-geographical materialist approaches to capitalist urbanization therefore provide important conceptual tools to explain how the commodification of naming rights contributes to the *tumultuousness* of urban landscapes within the context of neoliberal urban governance. Yet acknowledging the role of toponymic commodification as a strategy of neoliberal urbanism is only a starting point, rather than a closing argument, for theorizing the political economy of urban place naming.

Another theoretical framework that has considerable potential to critically analyze the geographies of naming rights is Bourdieu's (1990) theory of symbolic capital. The public recognition and visibility associated with the naming of places bestows a sense of prestige, or symbolic capital, upon the individuals or corporate entities recognized through place naming. According to symbolic capital theory, one form of capital can potentially be converted into other forms of capital (Bourdieu, 1986). The commodification of naming rights is a classic example of how the symbolic capital of a place name can be converted directly into economic capital, and, conversely, how economic capital can literally be used to purchase symbolic capital in the form of toponymic recognition.

Yet while naming rights may acquire economic value through the formal act of exchange, commodified toponymies only acquire symbolic value to the extent that officially-sponsored place names are *perceived* to have value among various "publics" and are thus incorporated into the spatial practices of everyday life. If a naming right is ignored or outright resisted, its symbolic value may be diminished irrespective of its officially-recognized legal status. The owners of naming rights – as well as the accompanying industry of naming rights specialists, consultants, etc. – therefore have a vested interest in promoting the legitimacy of toponymic commodification and the popular use of commodified toponyms on official signage and maps as well as in the media and among ordinary urban residents.

One approach that can be drawn upon to critically examine these processes of toponymic legitimation is hegemony theory. Vuolteenaho and Puzey (2018) use Gramscian hegemony theory to consider the reciprocity of power relations within the context of street naming and offer a model to develop an understanding of the circumstances under which people tend to accept commodified naming rights as an urban namescape's "new normal" (cf. Harvey, 2012, pp. 156–157). Hegemonic power operates through the production of particular socio-spatial relations as the taken-for-granted order of everyday life in order to govern through consent rather than force alone. The selling of naming rights is often framed by its promoters as a commonsensical practice that aligns with the prevailing wisdom of entrepreneurial governance in the neoliberal city, and it can therefore be viewed as an attempt to assert hegemonic power over urban place naming.

However, resistance to toponymic commodification does occur, and if we follow Laclau (2018 [2005]) in defining the act of naming as constituting "a highly cathected rallying point" (p. 231), this can help us make sense of what mobilizes people to individually or collectively resist the commodification of naming rights and the corporatized urban spaces which it produces. For those who inhabit cities, the "in-built" or "scripted" effects of naming rights may not only result in perpetually eroding the honorific significance of urban place naming as well as the place-anchored use values of toponymy; such practices also result in technical-cartographic instability and spatio-cognitive evanescence (Boyd, 2000). One potential direction to come to terms with these effects could be to take a cue from Lefebvrian rhythmanalysis (Lefebvre, 2004). This approach could trace the trajectories, time-frames, and everyday repercussions of toponymic commodification by focusing on the *temporal pulses* of place (re)naming events (i.e. the initiation or expiration of naming rights agreements) across different time periods or urban jurisdictions, and how distinct *toponymic rhythms* enter the lived worlds – i.e. biographical life-cycles, spatio-temporal practices, and forms of toponymic attachment – of different urban populations.

The theoretical perspectives outlined above provide several useful entry points to developing critical theorizations of toponymic commodification, yet urban scholars have only begun to theorize the spatio-temporalities of urban naming rights. The articles in this special issue are an important step forward in extending our empirical knowledge and theoretical understanding of the political economy of toponymic commodification, and it is to these contributions that we now turn.

Contributions toward a critical geography of urban toponymic commodification

Given the lack of geographical scholarship on the political economy of urban naming rights, the studies in this special issue break new ground in contributing to the development of a critical geography of toponymic commodification. They do so using a variety of research methodologies – including quantitative comparative analysis, qualitative case studies, fieldwork, and digital media analysis – as well as different theoretical frameworks such as theories of cultural hegemony, symbolic capital, monopoly power, rent, scalar politics, urban policy mobilities, and techno-imperialism. The contributors examine the geographies of urban toponymic commodification in a diverse array of geographical locales. In particular, they consider case studies ranging from the selling of naming rights for sports arenas in European cities and metro stations in Dubai to the failed attempt at renaming a New Zealand/Aotearoa town as part of a neoliberal place branding strategy and the role of philanthropic naming in the "Facebookification" of San Francisco's gentrifying neighborhoods. The articles in this collection not only document the growing prevalence of urban naming rights globally but also demonstrate how different public and private actors have perceived the phenomenon of toponymic commodification in divergent ways.

The first three studies focus on the commodification of naming rights in the realm of professional sports within the European urban context. Sports arenas are major landmarks in many contemporary cities, and their public visibility in the urban landscape as well as through media coverage of sporting events has made them primary sites for naming rights sponsorships. Vuolteenaho, Wolny, and Puzey (2019) provide a systematic, comparative analysis of the spread of naming rights sponsorships for sports and entertainment venues in England and Wales, Finland, Germany, Italy, Norway, and Scotland. Using both quantitative and qualitative data on over 300 football (soccer) grounds and indoor arenas, they analyze the geographies of naming rights and tensions around them as "a particular 'cultural' aspect in the world-wide mobilization of entrepreneurial discourses" associated with neoliberal urbanism. More specifically, they document how an increasing number of sporting and entertainment facilities in European cities have become stages for naming rights sponsorship contracts over the last three decades. Overall, the authors estimate that a combined total of 38% of all football grounds and indoor arenas in their study sites have been named after sponsors (33.2% of football grounds and 46.1% of indoor arenas, respectively), with many older football grounds representing iconic landmarks and community symbols whose sponsorship-based renaming has often met local opposition. Additionally, the adoption of naming rights sponsorships has been anything but uniform across Europe due to locally and nationally differing economic and regulatory circumstances as well as variably intensive civic and fan contestations. For instance, the highest proportion of football grounds have been named after sponsors in Germany (74.3%) compared to a low of 8% in Italy. Another intriguing difference is that naming rights agreements tend to be tailored for considerably shorter periods in smaller, more impecunious, European urban settings than in bigger cities with premium infrastructures and stronger economic clout.

Medway, Warnaby, Gillooly, and Millington (2019) explore the "scalar tensions" that have arisen among football club owners and fan communities in relation to three corporate naming rights sponsorships for football stadia in the Greater Manchester Region of the UK. Drawing upon a series of interviews and on-site fieldwork, their study highlights the place branding strategies of club owners; how corporate names have been embedded into the urban landscape; and the contrasting responses that commodified naming has received among different fan bases. The findings suggest that fan resistance to toponymic commodification is strongest in cases where a sports stadium has "occupied the same location with the same name" for an extensive period of time, leading fans to develop an emotional attachment to its traditional, non-commodified place-identity. By contrast, when a football team relocates to a new stadium in a different location, fans will likely be "far less resistant to the imposition of a corporate toponym." This underscores the importance of understanding the relationality of place naming as well as the emotional geographies of place attachment as a scalar phenomenon. Following Hagen's (2011) call for more research on the intersections between the geographies of place naming and the politics of scale, Medway et al. convincingly argue that the corporate naming of sports arenas is a key process through which tensions over the scalar politics of place are currently unfolding as part of "the corporate creep of neoliberal intervention in the toponymic landscape of towns and cities."

Creţan (2019) likewise explores the conflicting responses of sports fans to the selling of naming rights, in this case for a football team named "Poli" in Timişoara, Romania, following the fall of communism. Although much of the literature on naming rights has focused on fixed capital in the form of urban infrastructure – sports arenas, convention centers, etc. – Creţan's study considers how the commodification of naming rights for a sports team's identity resulted in the fragmentation of urban fan communities and competing claims over the legitimate use of a name. To generate revenue, the local authorities in Timişoara sold the naming rights of a football team that had been known since it was established in 1921 by the name of Poli Timişoara (or Poli) to a wealthy investor, who renamed the team Fotbal Club Universitatea Timişoara (FCU Timişoara). A decade later, the naming rights were resold to another investor, who was subsequently convicted of using the team's financial accounts in a money laundering scheme. Due to the investor's financial mismanagement, the team acquired significant debt and was disbanded. Local authorities and corporate sponsors responded by creating a new team largely composed of players from a neighboring town, which they named Asociaţia Club Sportiv Politehnica Timişoara (ACS Poli). However, the majority of local fans supported another team, Asociaţia Sportivă Universitară Politehnica Timişoara (ASU Poli), since most of the original Poli players were on the latter team. Based on 120 interviews with fans from both Poli teams, Creţan examines how inequalities in political, economic, and social capital shape struggles over cultural hegemony, place identity, and community, and how the right to name provides a focal point around which such conflicts may coalesce. This study therefore serves as a useful reminder that the commodification of naming rights can have unintended consequences that may increase social divisiveness within local communities.

Despite encountering resistance in some cases, the selling of naming rights has become a key aspect of the commercialization of sports over the past three decades, and many urban policymakers have also embraced toponymic commodification as an entrepreneurial strategy of neoliberal urban governance. Rose-Redwood, Sotoudehnia, and Tretter (2019) draw upon 37 interviews, on-site fieldwork, and archival materials to provide a comparative analysis of urban naming rights programs in Dubai (United Arab Emirates) and Winnipeg (Canada). Extending Bourdieu's (1990) work on symbolic capital, they argue that urban toponymic commodification is a prime example of symbolic/economic capital transformations in which the symbolic distinction bestowed through the naming of public places is converted directly into economic capital in the form of what they call "naming rent." The authors observe that viewing the spatial identities of urban public places as rent-generating assets has the "effect of privileging the potential economic exchange value of urban toponyms over other potential non-market values associated with urban place naming." By comparing Dubai's Metro Naming Rights Initiative with the Sponsor Winnipeg Program, their study also highlights the variegated outcomes of urban naming rights programs – with Dubai's initiative far surpassing Winnipeg's program in terms of the sponsorship revenue generated. They maintain that scope and visibility are crucial to understanding these divergent outcomes, and they conclude that "the revenue generated from urban naming rights programs does not always meet the optimistic expectations of policymakers." While some urban residents in both Dubai and Winnipeg raised concerns about the crass commercialism of toponymic commodification, others were generally indifferent to the issue, leading Rose-Redwood et al. to suggest that commodified toponymies are increasingly becoming normalized as part of everyday urban landscapes.

Yet the normalization of toponymic branding is not always a seamless process, as Kearns and Lewis (2019) demonstrate by considering a failed attempt among business elites in New Zealand/Aotearoa to rename an entire town in order to enhance its international reputation through brand promotion. In 2016, local business leaders led a campaign to rename the town of Blenheim (Waiharakeke in Māori) as "Marlborough City" since the Marlborough region in which the town is located is internationally renowned for its wine industry. Kearns and Lewis critically examine media coverage of the renaming proposal and the contentious debate that ensued. They contend that the campaign to rename Blenheim was a neoliberal project of place-making that was based upon a "pervasive yet nebulous economic nationalism that seeks to yoke all local and national identity to enhancing export growth." The authors see the failure of this place branding scheme as indicative of the limits to neoliberal urbanism when it is strongly opposed by those seeking to defend the "heritage" of place or reclaim indigenous toponymies. The issues at stake, of course, depend upon the context in which place naming occurs, and Kearns and Lewis's study exemplifies how the politics of profit-driven place naming and branding practices are always place-specific and intertwined with existing power relations in a given society.

Toponymic inscription has long been employed as a political technology of dispossession, erasure, and the imperialist appropriation of territory, and McElroy (2019) illustrates how the selling of philanthropic naming rights to wealthy elites is currently implicated in the material and discursive dispossession of socially marginalized groups as "techno-imperialism" transforms the fabric of urban spaces and communities. As a case in point, McElroy focuses on the renaming of a public hospital in San Francisco as the "Zuckerberg San Francisco General Hospital and Trauma Center" in 2015 after Facebook's CEO Mark Zuckerberg donated $75 million toward renovating the hospital (while a significant sum of money, it was less than 10% of the total public funds provided for hospital renovations). McElroy argues that, when tied to the acquisition of naming rights for urban public places, philanthropic donations should be understood not as a "gift" but rather as "a self-interested promotional strategy." Based on extensive participatory action research with anti-eviction activists in San Francisco, McElroy links the renaming of "The General" as "The Zuckerberg" with the broader politics of tech-driven gentrification in the city. This latter process of urban redevelopment has resulted in the eviction of long-time local residents and contributed to negative health outcomes for those who have been displaced (in some cases likely contributing to premature death due to the stress of relocation, particularly for low-income, elderly residents of color). McElroy highlights the irony of Zuckerberg's name being symbolically associated with promoting "health" in the city while Facebook and other tech companies are placing the health and wellbeing of vulnerable populations at further risk through displacement and dispossession.

Future research directions

From sports arenas to metro stations, hospitals, and other public places, the commodification of naming rights continues to remake urban cultural landscapes in the 21st century. The contributions to this special issue draw attention to the diverse ways in which toponymic commodification is transforming the identities of public places into time-limited, rent-generating commodities. By delving into the intricacies of particular case studies of urban naming rights and toponymic branding, these articles make a significant contribution toward developing a more in-depth understanding of naming rights as an arena of political, economic, and cultural contestation over the production of urban space. Yet there is much work that remains to be done, and we conclude by proposing a number of potential avenues for future research on the geographies of urban toponymic commodification. Among other topics of scholarly concern, future research might address the following thematic areas:

(1) *Institutional ethnographies of the naming rights sponsorship industry.* The selling of naming rights is a multi-million dollar industry, and there are numerous consultancy firms that promote and facilitate the sale of naming rights globally. The naming rights industry also conducts its own marketing research and holds conferences and workshops where "best practices" are circulated among those in the industry as well as potential clients in government, the corporate world, and the non-profit sector. Conducting institutional ethnographies of particular naming rights firms or the industry as a whole – through participant observation at conferences, workshops, and other industry-related events; interviews with key actors; and critically examining industry magazines, newsletters, and other publications – could offer new insights into the processes driving the naming rights agenda.
(2) *Urban policy mobilities and the uptake of naming rights among urban policymakers.* As naming rights policies become more widespread, future research should examine how knowledges of naming rights policies and practices circulate among urban policymakers. Such studies could also consider what leads some governmental entities to embrace naming rights policies while others reject such policies.
(3) *Reception and resistance to toponymic commodification.* Few studies have considered how urban residents, sports fans, etc., perceive naming rights sponsorships in different geographical locales and contexts. A number of the studies in this special issue address this matter, but further research is needed to better understand the reception of commodified toponymies in everyday speech and the extent to which there have been movements to resist the commodification of public place-identities through forms of toponymic activism. This can build upon the extensive body of research undertaken within the first wave of critical place name studies, which has focused on resistance to, and contestation of, practices of (re)naming urban places.
(4) *Critical quantitative analyses of naming rights policy outcomes.* How successful have urban naming rights policies been on their own terms with respect to net revenue generation in different urban contexts from small towns to large cities? This pragmatic question could serve as the basis of an applied research agenda that uses the techniques of economic calculation to potentially challenge the self-promotional claims of the naming rights industry and over-confident urban politicians who are convinced that naming rights sponsorships will generate vast sums of revenue for urban infrastructure.

(5) *Addressing power disparities in the uneven development of urban naming rights.* There is a need to develop new ways to theorize the power relations associated with the accumulation of economic and symbolic capital through the commodification of naming rights as well as the uneven development of naming rights geographies across different spatial and temporal scales. Although it is challenging to map the uneven geographies of naming rights markets globally, it is clear that the sums involved in naming rights transactions in the peripheries and semi-peripheries of the world economy pale in comparison to their counterparts nearer its hubs. Only fleeting mention of these power disparities has been made in the scholarly literature on toponymic commodification, and this issue thus deserves further consideration. The same notion applies to the elevated status of English as the lingua franca of late-modern, advanced capitalism (Heller, 2010). In many contexts with a subaltern position on the global map of linguistic power relations, commercial toponymies adopted or transmogrified from the Anglophone world are a key instrument to mobilize symbolic capital (Yurchak, 2000). Meanwhile, these re-inscriptions of a global hegemony are seductive for some locals but irksome for others, all of which contributes to urban naming rights serving as a contested emblem of our neoliberal times.

The themes outlined here are by no means exhaustive of the potential directions of future research on the geographies of urban naming rights. Yet regardless of what issues are considered, it is high time for geographers and other scholars to critically engage with the spatial politics of urban toponymic commodification, and we hope that the current special issue will encourage other urban researchers to join us in critically analyzing toponymic commodification as one among many political technologies of neoliberal urbanism.

Disclosure statement

No potential conflict of interest was reported by the authors.

ORCID

Reuben Rose-Redwood http://orcid.org/0000-0001-6384-5877
Craig Young http://orcid.org/0000-0002-4324-1483

References

Alderman, Derek. (2002). Street names as memorial arenas: The reputational politics of commemorating Martin Luther King in a Georgia county. *Historical Geography*, *30*, 99–120.

Alderman, Derek. (2008). Place, naming, and the interpretation of cultural landscapes. In Brian Graham & Peter Howard (Eds.), *The Ashgate research companion to heritage and identity* (pp. 195–213). Aldershot: Ashgate.

Alderman, Derek, & Inwood, Joshua. (2013). Street naming and the politics of belonging: Spatial injustices in the toponymic commemoration of Martin Luther King Jr. *Social & Cultural Geography*, *14*(2), 211–233.

Azaryahu, Maoz. (1996). The power of commemorative street names. *Environment and Planning D: Society and Space*, *14*(3), 311–330.

Berg, Lawrence. (2011). Banal naming, neoliberalism, and landscapes of dispossession. *ACME: an International Journal for Critical Geographies*, *10*(1), 13–22.
Berg, Lawrence, & Kearns, Robin. (1996). Naming as norming: 'Race', gender, and the identity politics of naming places in Aotearoa/New Zealand. *Environment and Planning D: Society and Space*, *14*(1), 99–122.
Berg, Lawrence, & Vuolteenaho, Jani (Eds.). (2009). *Critical toponymies: The contested politics of place naming*. Burlington, VT: Ashgate.
Bigon, Liora. (2009). Urban planning, colonial doctrines and street naming in French Dakar and British Lagos, c. 1850–1930. *Urban History*, *36*(3), 426–448.
Bourdieu, Pierre. (1986). The forms of capital. In John Richardson (Ed.), *Handbook of theory and research for the sociology of education* (pp. 241–258). New York, NY: Greenwood Press.
Bourdieu, Pierre. (1990). *In other words: Essays towards a reflexive sociology*. Stanford: Stanford University Press.
Boyd, Josh. (2000). Selling home: Corporate stadium names and the destruction of commemoration. *Journal of Applied Communication Research*, *28*(4), 330–346.
Bridge, Sam. (2018, November 16). Riyadh metro naming rights auction raises $278m. *Arabian Business*. Retrieved from https://www.arabianbusiness.com/transport/408244-riyadh-metro-naming-rights-auction-raises-278m
Burton, Terry. (2008). *Naming rights: Legacy gifts and corporate money*. Hoboken, NJ: Wiley.
Cohen, Saul, & Kliot, Nurit. (1992). Place names in Israel's ideological struggle over the administered territories. *Annals of the Association of American Geographers*, *82*(4), 653–680.
Creţan, Remus. (2019). Who owns the name? Fandom, social inequalities and the contested renaming of a football club in Timişoara, Romania. *Urban Geography*. doi: 10.1080/02723638.2018.1472444
Duminy, James. (2014). Street renaming, symbolic capital, and resistance in Durban, South Africa. *Environment and Planning D: Society and Space*, *32*(2), 310–328.
Eddy, Terry. (2014). Measuring effects of naming-rights sponsorships on college football fans' purchasing intentions. *Sport Management Review*, *17*(3), 362–375.
Gillooly, Leah, & Medway, Dominic. (2019). Sponsorships, stadia and naming rights. In Simon Chadwick, Daniel Parnell, Paul Widdop, & Christos Anagnostopoulos (Eds.), *Routledge handbook of football business and management* (pp. 199–208). London: Routledge.
Giraut, Frédéric, & Houssay-Holzschuch, Myriam. (2016). Place naming as dispositif: Toward a theoretical framework. *Geopolitics*, *21*(1), 1–21.
Hagen, Joshua. (2011). Theorizing scale in critical place-name studies. *ACME: an International Journal for Critical Geographies*, *10*(1), 23–27.
Hakala, Ulla, Sjöblom, Paula, & Kantola, Satu-Paivi. (2015). Toponyms as carriers of heritage: Implications for place branding. *Journal of Product and Brand Management*, *24*(3), 263–275.
Harvey, David. (1985). The geopolitics of capitalism. In Derek Gregory & John Urry (Eds.), *Social relations and spatial structures* (pp. 128–163). London: Macmillan.
Harvey, David. (2012). *Rebel cities*. London: Verso.
Heller, Monica. (2010). The commodification of language. *Annual Review of Anthropology*, *39*, 101–114.
Karimi, Ali. (2016). Street fights: The commodification of place names in post-Taliban Kabul City. *Annals of the Association of American Geographers*, *106*(3), 738–753.
Kearns, Robin, & Lewis, Nicolas. (2019). City renaming as brand promotion: Exploring neoliberal projects and community resistance in New Zealand. *Urban Geography*. doi: 10.1080/02723638.2018.1472445
Knox, Paul. (1991). The restless urban landscape: Economic and sociocultural change and the transformation of metropolitan Washington, DC. *Annals of the Association of American Geographers*, *81*(2), 181–209.
Laclau, Ernesto. (2018 [2005]). *On populist reason*. London: Verso.
Lauermann, John. (2016). Municipal statecraft: Revisiting the geographies of the entrepreneurial city. *Progress in Human Geography*, *42*(2), 205–224.

Lefebvre, Henri. (2004). *Rhythmanalysis: Space, time and everyday life.* London: Continuum.

Light, Duncan. (2004). Street names in Bucharest, 1990–1997: Exploring the modern historical geographies of post-socialist change. *Journal of Historical Geography, 30*(1), 154–172.

Light, Duncan. (2014). Tourism and toponymy: Commodifying and consuming place names. *Tourism Geographies, 16*(1), 141–156.

Light, Duncan, & Young, Craig. (2014). Habit, memory and the persistence of socialist-era street names in post-socialist Bucharest, Romania. *Annals of the Association of American Geographers, 104*(3), 668–685.

Light, Duncan, & Young, Craig. (2015). Toponymy as commodity: Exploring the economic dimensions of urban place names. *International Journal of Urban and Regional Research, 39* (3), 435–450.

Madden, David. (2018). Pushed off the map: Toponymy and the politics of place in New York city. *Urban Studies, 55*(8), 1599–1614.

Masuda, Jeffrey, & Bookman, Sonia. (2018). Neighbourhood branding and the right to the city. *Progress in Human Geography, 42*(2), 165–182.

McElroy, Erin. (2019). Data, dispossession, and Facebook: Techno-imperialism and toponymy in gentrifying San Francisco. *Urban Geography*, 1–20. doi: 10.1080/02723638.2019.1591143

Medway, Dominic, Swanson, Kathryn, Neirotti, Lisa Delpy, Pasquinelli, Cecelia, & Zenker, Sebastian. (2015). Place branding: Are we wasting our time? Report of an AMA special session. *Journal of Place Management and Development, 8*(1), 63–68.

Medway, Dominic, & Warnaby, Gary. (2014). What in a name? Place branding and toponymic commodification. *Environment and Planning A*, *46*(1), 153–167.

Medway, Dominic, Warnaby, Gary, Gillooly, Leah, & Millington, Steve. (2019). Scalar tensions in urban toponymic inscription: The corporate (re)naming of football stadia. *Urban Geography*. doi: 10.1080/02723638.2018.1446585

Myers, Garth. (1996). Naming and placing the other: Power and the urban landscape in Zanzibar. *Tidjschrift Voor Economische En Sociale Geografie, 87*(3), 237–246.

Nelson, Laura. (2017, February 27). Transit officials backtrack on controversial policy to sell naming rights for metro stations. *Los Angeles Times*. Retrieved from https://www.latimes.com/local/lanow/la-me-ln-metro-naming-rights-20170227-story.html

Rose-Redwood, Reuben. (2008). From number to name: Symbolic capital, places of memory, and the politics of street renaming in New York city. *Social & Cultural Geography*, *9*(4), 432–452.

Rose-Redwood, Reuben. (2011). Rethinking the agenda of political toponymy. *ACME: an International Journal for Critical Geographies, 10*(1), 34–41.

Rose-Redwood, Reuben, Alderman, Derek, & Azaryahu, Maoz. (2018). The urban streetscape as political cosmos. In Reuben Rose-Redwood, Derek Alderman, & Maoz Azaryahu (Eds.), *The political life of urban streetscapes: Naming, politics, and place* (pp. 1–24). New York, NY: Routledge.

Rose-Redwood, Reuben, & Alderman, Derek. (2011). Critical interventions in political toponymy. *ACME: an International Journal for Critical Geographies, 10*(1), 1–6.

Rose-Redwood, Reuben, Alderman, Derek, & Azaryahu, Maoz. (2010). Geographies of toponymic inscription: New directions in critical place-name studies. *Progress in Human Geography, 34*(4), 453–470.

Rose-Redwood, Reuben, Sotoudehnia, Maral, & Tretter, Eliot. (2019). "Turn your brand into a destination": Toponymic commodification and the branding of place in Dubai and Winnipeg. *Urban Geography*. doi: 10.1080/02723638.2018.1511191

Scauzillo, Steve. (2016, December 7). LA County Metro seeing $$$$ with new opportunity to sell naming rights to rail lines, stations. *San Gabriel Valley Tribune*. Retrieved from https://www.sgvtribune.com/2016/12/07/la-county-metro-seeing-with-new-opportunity-to-sell-naming-rights-to-rail-lines-stations

Scauzillo, Steve. (2017, February 24). Metro rescinds policy to sell corporate naming rights to rail lines, stations. *San Gabriel Valley Tribune*. Retrieved from https://www.sgvtribune.com/2017/02/24/metro-rescinds-policy-to-sell-corporate-naming-rights-to-rail-lines-stations

Sotoudehnia, Maral. (2013). *"Turn your brand into a destination": City branding, naming rights, and the neoliberalization of Dubai, UAE* (Unpublished Masters dissertation). Department of Geography, University of Victoria.

Sotoudehnia, Maral. (2018). Toponymic checksum or flotsam? Recalculating Dubai's grid with Makani, "the smartest map in the world.". In Reuben Rose-Redwood, Derek Alderman, & Maoz Azaryahu (Eds.), *The political life of urban streetscapes: Naming, politics, and place* (pp. 290–308). London: Routledge.

Sotoudehnia, Maral, & Rose-Redwood, Reuben. (2019). "I am Burj Khalifa": Entrepreneurial urbanism, toponymic commodification, and the worlding of Dubai. *International Journal of Urban and Regional Research*. doi: 10.1111/1468-2427.12763

Vuolteenaho, Jani, & Ainiala, Terhi. (2009). Planning and revamping urban toponymy: Ideological alternations in the linguistic landscaping of Vuosaari suburb, eastern Helsinki. In Lawrence Berg & Jani Vuolteenaho (Eds.), *Critical toponymies: The contested politics of place naming* (pp. 227–251). Farnham: Ashgate.

Vuolteenaho, Jani, & Kolamo, Sami. (2012). Textually produced landscape spectacles? A Debordian reading of Finnish namescapes and English soccerscapes. In Jani Vuolteenaho, Andrew Newby, Lieven Ameel, & Maggie Scott (Eds.), *Language, space and power: Urban entanglements* (pp. 132–158). Helsinki: Helsinki Collegium for Advanced Studies.

Vuolteenaho, Jani, & Puzey, Guy. (2018). "Armed with an encyclopedia and an axe": The socialist and post-socialist street toponymy of East Berlin revisited through Gramsci. In Reuben Rose-Redwood, Derek Alderman, & Maoz Azaryahu (Eds.), *The political life of urban streetscapes: Naming, politics, and place* (pp. 74–97). New York, NY: Routledge.

Vuolteenaho, Jani, Wolny, Matthias, & Puzey, Guy. (2019). "This venue is brought to you by . . . ": The diffusion of sports and entertainment facility name sponsorship in urban Europe. *Urban Geography*. doi: 10.1080/02723638.2018.1446586

Wanjiru, Melissa Wangui, & Matsubara, Kosuke. (2017). Street toponymy and the decolonisation of the urban landscape in post-colonial Nairobi. *Journal of Cultural Geography*, *34*(1), 1–23.

Woisetschläger, David, Haselhoff, Vanessa, & Backhaus, Christof. (2014). Fans' resistance to naming right sponsorships: Why stadium names remain the same for fans. *European Journal of Marketing*, *48*(7/8), 1487–1510.

Wright, John Kirtland. (1929). The study of place names: Recent work and some possibilities. *Geographical Review*, *19*(1), 140–144.

Yeoh, Brenda. (1996). Street-naming and nation-building: Toponymic inscriptions of nationhood in Singapore. *Area*, *28*(3), 298–307.

Yurchak, Alexei. (2000). Privatize your name: Symbolic work in a post-Soviet linguistic market. *Journal of Sociolinguistics*, *4*(3), 406–434.

"This venue is brought to you by...": the diffusion of sports and entertainment facility name sponsorship in urban Europe

Jani Vuolteenaho, Matthias Wolny and Guy Puzey

ABSTRACT

Drawing on a database of 193 football (soccer) grounds and 115 indoor arenas, as well as press releases and media reports associated with them, this study charts the diffusion of sporting and entertainment facility name sponsorship across metropolises, cities, towns, and smaller settlements in six European contexts. Our results show the emergence of naming rights deals in the 1990s, their peak in the mid-2000s, and the current situation with a steadier growth of name sponsorship. Thus far, the corporate re-branding of venues has remained less prevalent in Italy, Norway, and Scotland than in England and Wales, Finland, and above all Germany. In financing newly built venues, however, the corporatized landmark language in focus has become a practically invariable part of local growth, austerity and (re)branding policies. Despite voices of resistance in all regions studied here, pressure towards the corporate renaming of even hereditary, communally endorsed football stadiums is increasingly being felt by municipal and private-sector venue owners.

Introduction

In *Ethics, Money and Sport*, Walsh and Giulianotti write that contemporary sporting venue and competition naming rights herald "a revolutionary attitude towards nomenclature almost as radical as the Bolshevik renaming of the ancient city of St Petersburg, or the Khmer Rouge's re-titling of city streets in Cambodia" (2007, p. 2). Recent decades have witnessed an unprecedented surge in venue naming and name-changing exercises for commercial ends, starting in cities of the United States, and subsequently spreading to practically all urbanized parts of the globe (see more generally on the commodification of urban place naming: Light & Young, 2014; Medway & Warnaby, 2014; Rose-Redwood, 2011; Vuolteenaho & Ainiala, 2009). Across Europe, too, an increasing number of sporting and entertainment facilities (and, more recently, other types of principally "public" urban infrastructures) have become stages for this "innovative" revenue-generating strategy. Naming rights have been sold for many hundreds of European venues, whether by local councils with shrinking finances, or by private teams and corporations attempting to stay competitive. Nevertheless, comparative analyses portraying and attempting to explain this phenomenon on a continental or sub-continental scale have remained rare to say the least (for a partial exception, see Bezold, 2013). This article will, therefore, seek to explore the diffusion of naming rights sponsorship across a set of European national contexts with variably resource-rich sporting leagues and different venue infrastructures. By combining quantitative and qualitative comparative methodologies, we will ask what kinds of diachronic trends and explanatory factors (for instance in terms of city size and venue capacity), as well as manifestations of neoliberal urbanism and civic opposition, have characterized the adoption of venue name sponsorship in England and Wales, Finland, Germany, Italy, Norway, and Scotland (see a detailed list of research questions in the methodology section below). In particular, the focus will be on the extent to which this business-oriented toponymic alternative has been implemented at football (or soccer, the biggest spectator sport by far in most European national contexts) grounds, and with indoor sporting and entertainment venues.

To bolster our empirical mapping of this largely uncharted terrain, in the following theoretical section we will contextualize the new urban "landmark language" (Viljamaa-Laakso, 1999) in question as an aspect of the neoliberal city, generally characterized by a heightened entrepreneurial ethos, place-marketing initiatives and proactiveness in capturing globally mobile investment flows. Often implemented in connection with one-off or continual sporting events, venue naming rights deals represent a particular type of sponsorship that has had its recent popularity boosted by a decline in public financing in the context of urban austerity. Meanwhile, acts of selling venue naming rights facilitate commodification-led tendencies of cultural-linguistic globalization and (pseudo-)anglicization, public-private partnership arrangements, and arguably also new types of transurban connectivities in terms of learning profit-driven policies. While expected gains in return for naming rights deals range from image enhancement and new revenue streams to increased investment prospects, earlier studies have indicated that levels of acceptance for the sale of naming rights have tended to vary considerably among locals, and especially fans. In Europe, football venue naming rights deals, in particular, have not infrequently faced "traditionalist" opposition (e.g. Church & Penny, 2013; Woisetschläger, Haselhoff, & Backhaus, 2014). Next, we analyze a database of over 300 football grounds and indoor arenas, firstly through nation-specific overviews and exemplifying vignettes, illuminating trajectories, peaks and other characteristics in how the naming rights phenomenon has spread in these settings. In the synthesis that follows the nation-specific snapshots, we highlight general European trends and local- and national-scale peculiarities in the (non-)diffusion of this particular form of commodification of urban space. In conclusion, we propose future research directions for urban place-name studies suggested by our findings.

A remedy for event cities in financial straits? Conceptualizing venue name sponsorship as a corporatized landmark language

The recent surge in naming rights deals can be seen as a particular "cultural" aspect in the world-wide mobilization of entrepreneurial discourses, as well as a dimension of event-led urbanization aimed at encouraging inward investment and facilitating the rebranding of cities and their landmark infrastructures (on entrepreneurial urbanism and event-led urban growth and branding policies, see e.g. Broudehoux, 2017; Gratton, Shibli, & Coleman, 2005; Harvey, 1989; Lauermann, 2016; Smith, 2016). In neoliberal event cities where spaces such as football stadiums and multi-purpose indoor arenas are perceived as key catalysts of urban growth, the financial and promotional "gravity" of urban place and event names has been increasingly recognized by transnational mobile capital and local policy-makers alike. While the consequent interest in selling naming rights is certainly not restricted to sports facilities, sport sponsorship has long been a realm where enormous sums of money circulate annually. In these markets, locally influential growth coalitions and marketing partnerships have emerged, weaving city-specific webs between corporations and sports teams, developers, investment companies and local authorities.

Globally, the impact of sponsorship on urban redevelopment has been most commonly analyzed in relation to the bidding, planning, implementation, and aftermath of the Olympic Games, FIFA World Cup and UEFA European Championship tournaments as well as other one-off urban mega-events (e.g. Broudehoux, 2017; Klauser, 2011). In the United States, by contrast, sport-associated regeneration strategies have normally focused on facilities built predominantly for domestic spectator sports (Gratton et al., 2005). Importantly in this regard, Giulianotti (2011) has noted that it is not only major international events hosted in state-of-the-art arenas of world-class metropolises, but also "more time-space diffuse sporting occasions" (p. 3294) associated with domestic tournaments (such as the English Premier League) that influence economic and social fortunes in the cities of North America, Europe and beyond. Indeed, it is not only stadiums of worldwide repute, with veritable potential for mobilizing sponsorships with select world-class corporations, that have been subject to corporate name sponsorship in recent decades (Bezold, 2013; Herstein & Berger, 2013).

From a transurban interconnectivity or diffusion perspective, critical theorists from Harvey (1989) to Klingmann (2007) have labeled the "serial repetition" of successful redevelopment and marketing models as a hallmark of contemporary entrepreneurial urbanism. The iterative practices of neoliberal urban reforms cover the creative borrowing of architectural-stylistic fashions, implementing redevelopment solutions, commodified standardization of many urban symbolic forms, and privileging the use in urban branding of select "world languages," especially English as the globally hegemonic lingua franca of business. Instances of such linguistic landscape transformations, with urban spaces increasingly (re)signified through "immediately recognizable" designations, can be seen across continental Europe, and elsewhere, as typified by Yurchak's (2000) semiotic reading of toponymic "westernization" in post-Soviet Russia. Similarly, Vuolteenaho and Kolamo (2012) have explored the blossoming of "overworked globalisms" in Finnish urban settlements, arguing that these "extroverted" and "spectacular" urban nomenclatures form a "semi-hegemonic" trend across the country's leisurescapes. Elsewhere in Europe too, many recent sales of naming rights have led to the proliferation of urban place names that often have "nothing whatsoever to do with local geographical circumstances" (Viljamaa-Laakso, 1999, p. 82).

Besides such cultural-linguistic borrowing eroding the heritage-bearing function of urban names, the surging sales of urban naming rights mirror the recent intensification in transurban "policy mobilities" (see e.g. McCann, 2013; Peck & Theodore, 2010). The procedural sharing of "best practices" in sporting venue and other spatial naming rights is spreading, although this type of "transurban interconnectivity" (Giulianotti, 2011) is largely overlooked or only tangentially tackled in theoretical urban literature. Contemporarily, web portals (e.g. http://www.uksponsorship.com) efficiently deliver information about sport-, culture- and education-related sponsorship opportunities, while marketing consultancy firms laud their expertise in tailoring mutually beneficial contracts, providing bold instructions to stakeholders. According to SponsorPitch (2011), for instance, sponsors are guided to stick to short venue names ("the return on investment in terms of media impressions alone will be enormous" if this rule is followed), eschew renaming existing stadiums with heritage, and seek long-term contracts which show their commitment and hence help to gain the admiration and respect of fans and the media.

When seen as a distinct marketing communication method, sponsorship refers to a strategically planned way of donating resources, carried out in a way that is mutually beneficial to both sponsor and sponsee (Demir & Söderman, 2015). In the case of sporting and entertainment venues, the acquisition of naming rights usually translates into the sponsor's increased visibility among targeted audiences, especially through the new place brand's recurring evocations in the digital sphere and at the venue itself. Local venue owners may likewise benefit from the repute of the corporate brand after which their "homes" are named but, above all, they gain revenue in return for sold venue names. By supplementing owners' (or tenants') income streams from ticket and merchandise sales, in-stadium and kit advertising, hospitality services and the like, venue name sponsorship amounts to "a maximization of the economic utility of stadium space" (Gaffney, 2008, p. 205). Moreover, with the gradual institutionalization of corporatized urban place-naming practices, many publicly owned sporting venues have been also made accessible to naming rights deals. All this has increased the importance of name sponsorship in financing redevelopment projects in neoliberal event cities.

In this article, venue name sponsorship is conceptualized as a comparatively novel yet contested and potentially risky method of garnering revenue. In European cities and towns of different sizes and economic circumstances, local actors with varying economic clout have resorted to this funding source. Arguably, this tendency has been accelerated by the contemporary urban context of austerity policies and declining public subsidies (Schönig & Schipper, 2016), along with the simultaneous threat of bankruptcy hanging over many European sporting teams (see Kennedy & Kennedy, 2012). This article's empirical analyses will, however, demonstrate that not all recently built or renovated venues have become name-sealed hubs of brand alliances in Europe. In certain national contexts mapped below, recalcitrance towards the selling of naming rights at older venues has remained the rule rather than the exception. While the attachment felt by fans to stadium names is not dissimilar to the notion of toponymic attachment elaborated by Kostanski (2009), it is also notable that a number of owners of "stadiums with heritage" have learnt to value their inherited names as community assets, at least for the time being. This notion applies particularly to European football, a game paradoxically known both for its heavyweight commercialism and as a realm of sometimes enthusiastically traditionalist fan cultures in which a nostalgic drive toward the "moral economy of space and community" reigns supreme (Kennedy & Kennedy,

2010, p. 567; see also Bale, 2000; Bouchet, Hillairet, & Bodet, 2013; Hognestad, 2015; cf. in North America: Gerke, 2017).

Methodology

As Bezold (2013) notes, very limited empirical work has been thus far published on the European venue name sponsorship market. Apart from largely inaccessible marketing research and existing annual listings of top and select smaller venues with sponsored names, to our knowledge no prior systematic quantitative comparative studies exist on this topic in human geographic or urban studies literature, which is somewhat surprising given the recent prevalence of this phenomenon. That said, in order to start charting and explaining the geographical diffusion of naming rights, the new phenomenon's dynamic and fuzzy nature must be acknowledged, as several economic, business-specific (or sport-specific) and cultural factors can influence the adoption (or lack) of venue name sponsorship in specific urban and national settings.

For this study, we opted to compare the type of toponymic (re)branding under investigation from two overlapping perspectives. Firstly, to mirror the heterogeneity of economic, demographic, policy-related and cultural circumstances prevailing in European professional sports, we were keen to compare the diffusion of naming rights through a sample of geographical settings from continental, insular, and Nordic contexts, including some known for their world-leading national leagues, and others with less commodified leagues and less global media attention. To obtain an adequately rich sample, we chose to conduct the study in England and Wales (studying both countries together due to their integrated top football leagues), Finland, Germany, Italy, Norway, and Scotland. Secondly, we took a cue from Crawford's (2004, pp. 80–81) division between football, as a sport where venues regularly display a high degree of tradition-awareness, and "newer" organized sports that "have been successfully stripped of their (relatively) little history and sold as easily accessible entertainment events." By comparing football grounds with indoor event arenas (more commonly also used for non-sport occasions and show-business), we thus aimed at a corpus reflecting differences in the propensity of adopting market-led changes and trading off tradition in favor of corporate toponymy.

In each case, we collected a subset of data comprised of the home football grounds of the clubs in the top two national leagues in the 2016–17 season (or the 2017 season in Finland and Norway), as well as the national football stadium (where applicable). Football-related data were then supplemented by available statistics on the 15–20 largest indoor arenas – at least occasionally used for sporting events – in each geographical context by spectator capacity.

In operationalizing the explorative statistical objectives of this research, we settled on four dependent variables intended to measure the propensity for naming rights in the respective countries and localities:

- Venues currently with a sponsored name or a non-sport corporate owner's name (yes/no);

- Venues that have had a sponsored name or a non-sport corporate venue owner's name (yes/no);
- The year of each venue's first naming rights contract (where applicable);
- The number of separate naming rights deals per venue.

In terms of available independent variables explaining the diffusion of name sponsorship, it often proved difficult to obtain adequate and consistent information about some potentially useful characteristics of venues and naming rights deals. For instance, information on the monetary sums involved in naming rights deals, or on venues' public/private ownership patterns, were not always publicly traceable, being often seen as commercially sensitive. Measuring and categorizing the linguistic-semantic features of the venue names or the effects of venues' intra-urban locations on the incidence of naming rights in an adequately rigid fashion are other tasks that fell outside the scope of this study, but which offer much scope for future work. However, for nearly all of the 308 sporting venues included in the dataset, we were able to trace systematically the facilities' original year of construction, the business sector of venue name sponsors, and three other independent variables measuring their size and status:

- Capacity (maximum spectators for sporting events);
- Local urban population;
- Key tenant's position in the hierarchy of national leagues (first league/second or lower league/no key tenant).

The analysis involved two phases, intended to answer two sets of research questions. In the first phase, the focus was on basic statistical analyses (frequencies, measures, cross-tabulations and chi square tests) of the aforementioned variables. On the basis of the collected statistical data, we asked:

(1) How do different independent variables explain the adoption and non-adoption of naming rights deals in these European countries? For instance, what types of venues have been national trend-setters, or more adverse to entering such deals?

In the second phase we adopted a more qualitative approach. We interpreted our results through a broader lens of statistical data, content analyses of associated media sources, and dialogues with theoretical literature, aiming to answer the following research questions:

(2) How are the commodification-related processes and realities of globalization, revenue-generation and austerity reflected in the European data?
(3) What kinds of diachronic trends and peaks as well as learning procedures have characterized the spread of name sponsorship in Europe?
(4) How prevalent is resistance to name sponsorship in Europe, and what kinds of factors explain these critical public responses?

National snapshots of venue naming rights

England and Wales

In our English and Welsh data, 15 football grounds (33.3%) and 13 indoor arenas (65.0%) have at some point had names related to non-sports corporations and brands (see Table 1). The current figures are moderately lower: 26.7% and 55.0% respectively, indicating a weaker counter-trend away from the sale of naming rights.

In our sample, a limited number of naming rights deals were already made in the 1990s. Symptomatic of the turbulence of sports and entertainment businesses in these countries, the five deals made before 2000 were short-lived, with three of these venues having changed names twice to date. The pioneering facility in football was in Huddersfield: the *McAlpine Stadium* (1994–2004), later renamed the *Galpharm Stadium* (2004–12) and the *John Smith's Stadium* (2012–). The earliest sponsored name for an indoor arena was *NYNEX Arena* (1995–98), subsequently re-baptized as the *Manchester Evening News Arena* (1998–2011) before turning to its current non-corporate name *Manchester Arena*, with an interlude when it was known as the *Phones 4u Arena* (2013–15).

Still, the first real boom in venue name sponsorship in England and Wales began in the 2000s, when six indoor arenas and eight football grounds were branded after non-sports corporate sponsors, as exemplified by *Echo Arena Liverpool* and Leicester City's *Walkers Stadium* (since 2011 the *King Power Stadium*), both built on former industrial sites. At the upper end of the economic spectrum, in 2004, a Dubai-based airline committed to pay Arsenal F.C. £100 million for 8 years of shirt sponsorship and 15 years of naming rights for what would become the *Emirates Stadium* ("Arsenal name new ground," 2004).[1] This then record-breaking sum represented one of the first such deals for a true European mega-venue, along with the *Allianz Arena* in Munich, named after a German insurance giant in 2005, and London's entertainment complex *The O2*, which also took on its corporate name in 2005.

More recently, the English and Welsh corpus suggests contradictory signals about the trend's continuing diffusion. For football stadiums, in particular, the growth rate in naming rights has at least temporarily decelerated in the 2010s. One likely explanation is linked with what Russell (2006, p. 21) has dubbed the "turn to history and commemoration," yielding a proliferation of genuflections to the past in the otherwise "hyper-commodified" world of British, and especially English, football (cf. Giulianotti, 2002). This is a peculiar feature in comparison to many other European countries. Even English football's long-standing flagship enterprise Manchester United F.C. has repeatedly ruled out renaming its home, with commercial director Andy Anson stating: "Old Trafford will always be Old Trafford, we are not going to do a naming rights deal" (as cited in "Man Utd rule out renaming stadium," 2005). Explicit fan pressure against naming rights has been documented in previous studies on English football culture, with probably the most widely known example being ardent protests by Newcastle United F.C. fans in 2009–12 against the club owner Mike Ashley's stadium name sponsorship arrangements for St James' Park (opened in 1892) (Cleland & Dixon, 2015).

Finland

In percentage terms, our Finnish corpus appears similar to England and Wales, with 7 football grounds (33.3%) and 14 indoor arenas (70.0%) having had sponsored names. Finland is the only country in our study where the average size of indoor arenas trumps football grounds (see Table 1). This is primarily due to the status of ice hockey as Finland's biggest spectator sport.

When naming rights as an urban landmark language first emerged in Finland, new multi-purpose arenas were in the vanguard, marketed at the time with elitist overtones (Viljamaa-Laakso, 1999). An early forerunner, even in the broader European context, was the indoor arena *Typhoon* in the regional capital Turku, which took on that name in 1990 in connection with the regional banking institution TYP, and has since been known as the *Elysée Arena, HK Areena,* and *Gatorade Center*, as well as *Turkuhalli* in periods without sponsorship. This was followed by other indoor arenas in the metropolitan area of Helsinki such as *Hartwall Areena* in 1997, sponsored by a drinks company, and *LänsiAuto Areena* in 1999, sponsored by a local car retailer.[2] In the 2000s, the trend for naming deals accelerated, increasingly involving less "upmarket" and often relatively old, middle-sized indoor arenas (built during the heyday of nationwide, largely publicly funded sporting infrastructure construction between the mid-1960s and mid-1980s) in provincial towns whose ice hockey teams tended to perform in the major national ice hockey league. As with many other regions peripheral to the sport and entertainment industries' capital flows, name sponsors in Finnish sportscapes are characteristically national or regional firms, institutions or publicly owned companies, and only very exceptionally global corporate giants or brands (Vuolteenaho & Kolamo, 2012).

Given that many pre-1990 venues have seen naming rights sold for short-term economic gains, it is striking that cultural opposition towards this swiftly spreading phenomenon has remained comparatively weak in Finland. There have been occasional signs of civic disaffection and institutional resistance, with suspicions aired in newspapers and online forums, the Finnish Broadcasting Company's 2011 decision to boycott sponsored venue names in its broadcasts and reports, and the Helsinki Olympic Stadium Foundation announcing that it will not sell the sacralized name it owns. Otherwise, the heritage value of names has rarely counted in the revenue-prioritizing deliberations surrounding naming deals. Rather, in the pragmatic mindsets of many politicians, authorities or marketers, particularly in smaller provincial municipalities, the key challenge in the name-sponsorship market concerns the difficulty of trading sports facilities' names, as evidenced by complaints in our media material of how "there are no buyers queuing" (cited in Korkala, 2006).

Germany

The highest density of name sponsorship among the countries covered here is in Germany, where 35 of the venues considered (63.6%) are currently sponsor-named. If focusing only on football grounds, the percentage rises to 65.7%, and if football grounds that have previously had sponsored names are included, the percentage rises further to

Table 1. Key statistical data.

	Venue type	Total number	Has been named after a sponsor	Currently named after a sponsor	Minimum capacity	Maximum capacity	Average capacity	Earliest year of construc-tion	Percentage built after 2000
ENGLAND/WALES	Football grounds	45	15 (33.3%)	12 (26.7%)	6,912	90,000	33,691	1877	26.7%
	Indoor arenas	20	13 (65.0%)	11 (55.0%)	5,500	21,000	10,889	1871	45.0%
FINLAND	Football grounds	21	7 (33.3%)	7 (33.3%)	500	39,784	5,356	1926	9.5%
	Indoor arenas	20	14 (70.0%)	13 (65.0%)	3,500	13,349	6,081	1965	10.0%
GERMANY	Football grounds	35	26 (74.3%)	23 (65.7%)	13,138	81,360	38,186	1910	34.3%
	Indoor arenas	20	12 (60.0%)	12 (60.0%)	6,000	18,700	12,314	1909	40.0%
ITALY	Football grounds	37	3 (8.1%)	3 (8.1%)	5,510	81,277	25,434	1911	2.7%
	Indoor arenas	20	7 (35.0%)	7 (35.0%)	5,500	16,600	8,146	1956	25.0%
NORWAY	Football grounds	32	8 (25.0%)	8 (25.0%)	1,265	27,200	8,004	1919	25.0%
	Indoor arenas	20	4 (20.0%)	4 (20.0%)	3,500	15,000	6,033	1963	30.0%
SCOTLAND	Football grounds	23	5 (21.7%)*	5 (21.7%)	2,020	60,411	16,893	1879	17.4%
	Indoor arenas	15	3 (20.0%)	3 (20.0%)	1,100	13,000	5,175	1938	26.7%
	Football grounds	193	64 (33.2%)	58 (30.1%)	500	90,000	23,579	1877	20.2%
COMBINED	Indoor arenas	115	53 (46.1%)	50 (43.5%)	1,100	21,000	8,234	1871	29.6%
	Overall	308	117 (38.0%)	108 (35.1%)	500	90,000	17,850	1871	23.7%

* There is also one case in the Scottish corpus of a football ground (Ibrox Stadium) where naming rights have been sold, but which has never been named after a sponsor. This additional naming rights deal is accounted for in Figures 1 and 2.

74.3%. As for the German indoor arenas in our sample, 60.0% currently carry the name of a sponsor, a value second only to Finland.

The first case from the German sample of a sponsor-named football ground dates from 1997, when the *Sportpark Ronhof* in the city of Fürth in Bavaria was renamed the *Playmobil-Stadion*, after the stadium's then owner, the toy manufacturer Playmobil. Subsequently, there was a peak in football-related naming rights deals in Germany in the first decade of this millennium, in line with the construction or modernization of many of the country's biggest football grounds in preparation for the 2006 FIFA World Cup. This wave of development had a significant impact on football in Germany, with even many cities that were not among the hosts of World Cup games building or refurbishing stadiums to keep up with state-of-the-art infrastructure elsewhere. Naming rights for many of these facilities were sold, with sponsors playing a key role in many projects, either directly contributing to finance construction, or re-financing the work when purchasing naming rights at a later date. In fact, 11 of the 12 stadiums built from 2000 onwards bear the name of a sponsor.[3] Since the start of this modernization boom, the economic mantra of German professional football appears to have become that in order to be successful on the pitch, a team first needs state-of-the-art infrastructure.

According to Bezold (2013), "generally fans in Germany have a very pragmatic understanding of accepting new sponsoring tools like naming rights as an additional source of income for the football club although there are cases of opposition" (p. 129). One exception occurred when the first two stadium naming rights deals brokered by officials at 1. FC Nürnberg resulted in the names *easyCredit-Stadion* (2006–12) and *Grundig Stadion* (2013–16). After the naming rights were sold for a third time in 2016 and acquired by Consorsbank, club supporters reached an agreement to rename the stadium as the *Max-Morlock-Stadion* (commemorating a legendary player) if fans contributed €800,000 to the naming rights deal. In the spring of 2017, a crowdfunding initiative to support this collected less than half of the required amount, but Consorsbank agreed to cover the remainder, making the *Max-Morlock-Stadion* a visible urban reality for the start of the 2017–18 season ("1. FC Nürnberg spielt künftig im Max-Morlock-Stadion," 2017). Quite exceptionally in the German context, anti-corporate FC St. Pauli supporters in Hamburg have also insisted that "the name Millerntor-Stadium cannot be sold, modified or changed for the purpose of advertising, sponsoring or as a reward for any financial contributions to the club" (as cited in Merkel, 2012, p. 367; see also Bezold, 2013).

Italy

Compared to the other countries considered here, Italy exhibits a very different situation in the diffusion of sponsor-named sports venues. While the statistics for indoor arenas that have been sponsor-named (35.0%) are not far off the average percentage for indoor arenas across the whole study (46.1%), the percentage of football grounds carrying a sponsor name is by far the lowest (8.1%). By the end of the 2016–2017 season, this corresponded to only 3 relatively provincial venues out of 37 grounds in the top two divisions of Italian football: the *Mapei Stadium – Città del Tricolore* in Reggio Emilia; the *Orogel Stadium – Dino Manuzzi* in Cesena; and the *Dacia Arena* in Udine.

The stadium in Reggio Emilia represents a forerunner in Italian naming rights, having already borne the name *Stadio Giglio* (after a local dairy product manufacturer) from 1995 to 2012, when the privately built ground was used by A.C. Reggiana 1919, the city's biggest club, now relegated to the lower leagues. At the end of the sponsorship deal, its name was changed to *Stadio Città del Tricolore*, referencing the city's role in the development of the Italian flag. In 2013, U.S. Sassuolo Calcio, a team from a town near Reggio Emilia, was promoted to Serie A, and started using the stadium for its home games. In this process, the stadium was renamed as *Mapei Stadium – Città del Tricolore* after the chemical company chaired by the club owner.

It is striking that, while the sponsorship-based (re-)branding of indoor arenas in Italy has generally followed the broader European pattern, sales of football-related naming rights in this major European country with such a globally followed football tradition have not yet come close to the levels seen in much smaller countries like Finland, Norway or Scotland. One can speculate that this situation may stem from Italy's mainly publicly owned and older stadium building stock. Moreover, the existence of ultra-fanatic and often politically extremist supporter groups, hooliganism and organized crime in and around the stadiums, low levels of commercial "professionalization" in many clubs' management, and various financial problems among clubs may have made the purchasing of venue naming rights a less attractive option for potential sponsors with a predilection for a family-friendly and polished brand image.

Besides the three sponsor-named stadiums mentioned above, another special case is *Juventus Stadium* opened in 2011 as by far the most modern football venue in Italy. This stadium in Turin, a more metropolitan location, was renamed in July 2017 as *Allianz Stadium*, after the aforementioned Munich-headquartered insurance company that also lends its name to Germany's second-largest stadium, the *Allianz Arena*. Time will tell whether this conspicuous six-year contract will trigger a boom in naming rights in other Italian cities in the coming years.

Norway

Of the football grounds in our Norwegian data, only 8 (25.0%) presently have sponsored names. More strikingly, the 4 indoor arenas with sponsored names represent an even smaller percentage (20.0%), which is considerably lower than in the other national contexts in this study, apart from Scotland. This relatively limited number of venue name sponsorships in Norway may be explained by sometimes ardent fan opposition against the commodification of football events (see Hognestad, 2015), but also by recalcitrant attitudes among authorities, as well as economic prosperity which may make seeking name sponsors a less compelling imperative or austerity-easing tool for Norwegian teams and municipalities.

The earliest example of a naming rights deal within the Norwegian corpus is from 2002, with *Color Line Stadion* in Ålesund, a medium-sized town in the western county of Møre og Romsdal. This football stadium was opened in 2005, but the deal predates the construction of the stadium itself, as Olav Nils Sunde – the majority stakeholder and president of the Norwegian ferry operator Color Line – donated NOK 10 million towards construction, with naming rights apparently part of the deal from the outset ("Gir ny stadion til Aalesund," 2002). Color Line already had experience with naming deals from Germany: the indoor arena in Hamburg currently known as the *Barclaycard Arena* was originally called the *Color Line Arena*, having acquired that name a few months before the deal in Ålesund was announced.

Another case with documented transnational dimensions is in the town of Molde, also in Møre og Romsdal. Molde FK's current home ground opened in 1998 and was originally known as *Molde Nye Stadion* (literally "Molde New Stadium"), or simply *Molde stadion*. Construction had largely been funded by Kjell Inge Røkke, the chair and majority shareholder of Aker, a major Oslo-headquartered company involved in various maritime, construction and engineering activities, and his business partner Bjørn Rune Gjelsten. The pair are known, among other things, for their controversial 1997 purchase of the then English Premier League club Wimbledon F.C., which eventually relocated to Milton Keynes in 2003. *Molde stadion*, tellingly nicknamed *Røkke-løkka* ("Røkke's Paddock"), had its name officially changed in 2006 to *Aker Stadion*, in recognition of the long-term of the venue and its tenants by Aker.

Interestingly, the Aker Stadion renaming occurred at the height of an economic upswing, coinciding with a substantial number of new venue naming rights deals not only in Norway, but also in Germany, Finland, England and Wales. In Norway, at the tail end of this wave was the home ground of Sandefjord Fotball, which was named *Komplett.no Arena* in a deal with the online shopping company Komplett Group in 2007. The name was later simplified to *Komplett Arena*, but not before it had caused problems with the Norwegian Public Roads Administration refusing to put up signs to the stadium, as this could be construed as advertising (Rivrud & Lilleeng, 2008). Arguably a signal of institutional resistance among public authorities towards corporate-branded venue names, road signs in Sandefjord now state the name of the club instead of the stadium.

Scotland

Of the football grounds in our Scottish sample, 5 (21.7%) have had sponsored names. While this is less than the percentage for the grounds in the English and Welsh corpus, it is curious to observe that all the football grounds with sponsored names are at the smaller end of the spectrum, whereas the larger and typically older stadiums in Scotland carry tradition-bound names.[4] Meanwhile, the relatively modest number of large or middle-sized indoor arenas in Scotland – of which 3 (20.0%) have sponsored names – makes it difficult to make rigorous conclusions about these.

The earliest example in the Scottish dataset of a sponsored stadium name is from 2000, when the home ground of Dumbarton F.C. was opened with the name *Strathclyde Homes Stadium*. Dumbarton's ground is by some distance the smallest of all the Scottish stadiums in our corpus, with a capacity of 2,020, and has had four different sponsored names. This exemplifies a tendency among a substantial share of small venues, typically located outside metropolitan cores, that have already gone through multiple short-term renamings both in Scotland and elsewhere in Europe.

There is one particularly intriguing urban case in Scotland of a historic football ground, first opened in 1899, where naming rights have been sold but not used. Naming rights to *Ibrox Stadium* (formerly *Ibrox Park*), the Glasgow home ground of Rangers F.C. and third largest of Scottish football grounds, were sold in 2012 to the aforementioned billionaire Mike Ashley, the founder of Sports Direct and owner of Newcastle United F.C. (Jackson, 2014). This happened while Rangers were in considerable financial difficulty that had caused them to be forcibly relegated from the Premiership directly into the fourth-tier league. Around the time the deal is said to have already been made, an article was published on Rangers' official website with the club's chief executive apparently suggesting a naming deal may be close, and persuading supporters to endorse the prospective positive impact of any potential renaming on performances on the pitch:

> Charles Green says Rangers are close to striking a deal to sell the naming rights to Ibrox – but only on terms the fans are comfortable with. [...]
>
> "I'm sure people who have been associated with the club for longer than I have are still going to refer to its historical name. However, it is that history that attracts a sponsor. That's why people are prepared to pay for the naming rights of some stadia but not others. [...] When they did the naming rights deal at St James' Park, it brought cash in and that cash was used to acquire [striker] Demba Ba. He scored loads of goals then they sold him for £7 million so that shows what selling the naming rights can do. If you can use that money well, it'll bring further money in." (Dickson, 2013)

Two years later, shortly after the story emerged that the rights had been sold by Rangers for the token sum of £1, Ashley returned the unused naming rights, while also lending the club further money (McLaughlin, 2014). As shown again here, it is not entirely uncommon to see unexpected turns of events associated with naming rights deals, and even less so to see fans as a significant stakeholder group in relation to such deals.

Interpreting the diffusion of venue naming rights in urban Europe

To start to sum up key comparative characteristics in the adoption, or otherwise, of venue naming rights in the six national contexts detailed above, of the overall 308 venues, 117 (38.0%) carry or have carried the name of a sponsor (see Table 1). While the percentage among football grounds is slightly lower (33.2%) than the average, the percentage of sponsor-named indoor arenas is markedly higher (46.1%), a pattern reflecting the fact that the indoor arenas are on average more recently built. Another pivotal observation concerns the phases of the generally contemporaneous spread of the phenomenon in the urban landscapes analyzed across the continent. Figure 1 corroborates the previous section's narrative accounts by showing the emergence of early naming rights deals in the 1990s in each of the countries, except for Norway and Scotland. Since then, naming rights deals have spread into all the countries considered, becoming an established source of revenue for sports and entertainment businesses and local governments alike. In general, the spread of naming rights deals reached a peak in the first decade of the new millennium. Deals have continued to spread since 2010, albeit with temporarily reduced intensity (especially in Germany, and England and Wales, suggesting that the venue naming rights market may be approaching saturation in these countries).

Arguably, the popularity of naming rights deals varies between the studied European contexts according to several economic, regulatory, and more broadly cultural factors. The ranking of which countries have been trend-setters in corporate name diffusion, or

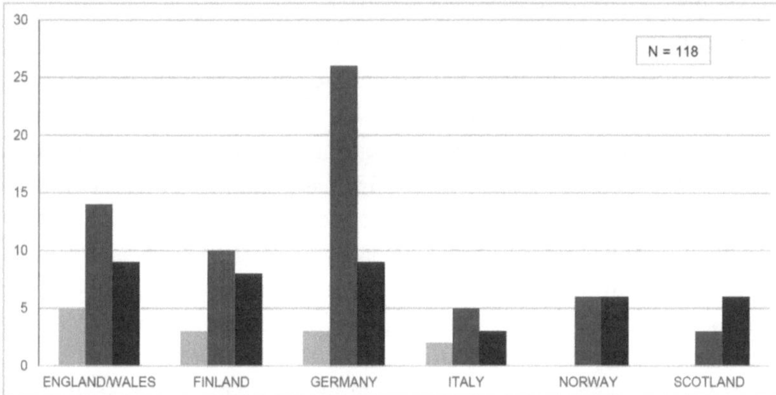

	ENGLAND/WALES	FINLAND	GERMANY	ITALY	NORWAY	SCOTLAND
First deal 1990–99	5 (3 + 2)	3 (0 + 3)	3 (2 + 1)	2 (1 + 1)	0	0
First deal 2000–09	14 (8 + 6)	10 (5 + 5)	26 (16 + 10)	5 (0 + 5)	6 (5 + 1)	3 (2 + 1)
First deal 2010 onwards	9 (4 + 5)	8 (2 + 6)	9 (8 + 1)	3 (2 + 1)	6 (3 + 3)	6 (4 + 2)

Figure 1. Year of first naming rights deals (football grounds + indoor arenas). The summary table shows absolute figures for football grounds and indoor arenas in brackets.

more adverse to such deals, proves relatively straightforward. At one extreme, Germany's sporting venues (69.1%), with both football grounds (74.3%) and indoor arenas (60.0%), are positioned at the top end of our corpus (see Table 1). Besides a nation-wide spate of large-scale infrastructural developments and subsequent sales of venue naming rights in the run-up to the 2006 FIFA World Cup, the strict financial regulations of the Deutsche Fußball Liga (German Football League) prevent clubs from spending more than they earn, steering them to squeeze out the maximum revenue possible from their venues. Meanwhile, the evolution of relationships between German fan groups critical of the game's commodification and the game's commercial forces towards more compromising attitudes, as analyzed by Merkel (2012), is likely to have paved the way for greater acceptance of venue naming rights sales in German football.

At another extreme, figures for Italy, Norway, and Scotland are considerably below the average share of sold venue naming rights. As mentioned, barely a handful of Italian football grounds are currently named after a sponsor or a company owning the facility. Disreputable phenomena associated with Italian football grounds may have limited their attraction as objects of sponsorship. In contrast, the comparative health of public finances in Norway and the relative age of the larger stadiums in Scotland (along with instances of resistance from fans in attempts to change venue names in both contexts), are among explanatory factors for the small number of naming rights in these countries. For their part, England, Wales, and Finland are positioned between the aforementioned extremes in the distribution of naming rights, with England and Wales characterized by the coupling of hypercommodified football with nostalgic venue branding motifs, and Finland by numerous corporate hockey arena names and a less vibrantly traditionalist culture of football fandom (Vuolteenaho & Kolamo, 2012).

As the above-quoted persuasive address of Glasgow Rangers fans by the club's chief executive implied, clear signs of financial pressure to sell naming rights can nonetheless be seen across Europe. Despite voices of resistance in all regions studied here, it seems this pressure is increasingly being felt at venues "with heritage," a tendency that stands in contradistinction to marketing experts' recommendations such as those given by SponsorPitch (2011). In England, for instance, plans for naming rights deals have been explored by several clubs that have nonetheless stuck with their traditional stadium names. As Wolverhampton F.C. chairman Steve Morgan put it in an interview, the club's home could well be renamed "the 'Something' Molineux Stadium. Emirates are paying Arsenal about £50million over 15 years and, if we get serious sponsorship, it would be idiotic not to consider it" (as cited in "Molineux could be renamed," 2008). Furthermore, while newly built middle-sized football grounds and bigger indoor arenas have most typically been national trend-setters in adopting this particular revenue-generating tool, there are currently signs of the phenomenon gaining traction among smaller venues and less resourceful lower-tier clubs (Bezold, 2013). In fact, multiple separate naming rights deals already figure relatively prominently in smaller venue categories in Finland, Italy, Norway, and Scotland (Figure 2).[5] Equally strikingly, our findings reveal that, in smaller European urban contexts, contracts often last only for a couple of years or even less, whereas average naming contracts in the U.S. span between 15 and 30 years (Bezold, 2013).

Furthermore, certain transurban interconnectivities are traceable on the basis of our quantitative and qualitative data. Firstly, the blossoming of "overworked globalisms" (Vuolteenaho & Kolamo, 2012) in European naming rights markets is most blatantly manifested through the prestige-enhancing generic *arena* (a kind of *sine qua non* for aspiring indoor venues, but also for a growing number of football grounds). Secondly, the re-appearance of identical or nearly identical corporate namesakes between different national and local contexts is an observable trend. Cases of such "name twins" include the *Mercedes-Benz Arena* in Berlin (formerly *O_2 World Berlin* in 2006–15) and the *Mercedes-Benz Arena* in Stuttgart, as well as the two *Barlaycard Arena*s in Birmingham and Hamburg, or Munich's *Allianz Arena* and Turin's newly re-branded *Allianz Stadium*. In essence, these examples epitomize a trend where heavyweight sponsors,

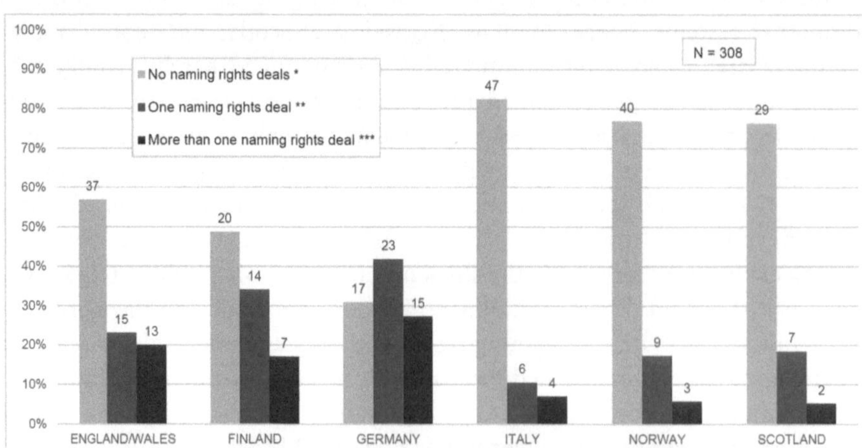

Capacity	England/Wales *	**	***	Finland *	**	***	Germany *	**	***
Less than 8,000	13.5%	20.0%	7.7%	90.0%	85.7%	71.4%	5.9%	4.3%	0.0%
8,000 – 20,000	10.8%	40.0%	38.5%	5.0%	14.3%	28.6%	47.1%	47.8%	40.0%
More than 20,000	75.7%	40.0%	53.8%	5.0%	0.0%	0.0%	47.1%	47.8%	60.0%

Capacity	Italy *	**	***	Norway *	**	***	Scotland *	**	***
Less than 8,000	25.5%	33.3%	25.0%	72.5%	55.6%	66.6%	37.9%	42.9%	50.0%
8,000 – 20,000	40.4%	33.3%	50.0%	22.5%	44.4%	33.3%	48.3%	42.9%	50.0%
More than 20,000	34.0%	33.3%	25.0%	5.0%	0.0%	0.0%	13.8%	14.3%	0.0%

Figure 2. Number of separate naming rights deals per venue (football grounds and indoor arenas), in total and by capacity.

with their global(izing) power brands (Hollands & Chatterton, 2003; Raento & Douglass, 2001), seek market credibility and expansion through brand alliances, "built corporate communication" and their own "signatures" inscribed onto different metropolitan landscapes (see e.g. Bielzer, 2013; Bouchet et al., 2013).[6] In more impecunious settings, such overlaps are not uncommon either, as evinced in Finland by *Oulun Energia Areena, Synergia-areena* and *Energia Areena*, all eponymous to municipality-owned energy companies. Among other instances of trans-locally operating venue name sponsorship markets are Mike Ashley's maverick moves to market his retail business empire Sports Direct around the U.K. (with short-lived naming rights blueprints or implementations in at least Newcastle, Glasgow, Oldham, and Blackpool), as well as the undertakings by the Norwegian owners of Aker and Color Line.

In the European urban context characterized by a patchwork of smaller and medium-sized cities alongside fewer true mega-cities (Pradel i Miquel, Paalzow, & Martin-Brelot, 2013), however, locally specific name sponsorship arrangements are more prevalent in the light of our data. Taking another example from Finland, the new corporate name *Trust Kapital Areena* for an ice hockey venue in Kuopio (until then bearing its original no-frills name *Kuopion jäähalli*), was announced in 2012. On this occasion, the tenant team Kalpa's chief executive dubbed a miscellaneous set of locally based corporations from brewing and forestry machinery to a local newspaper and banking companies as its "mega-partners," and specifically thanked the city council for its collaboration in the forging of the naming rights deal (Kalpa Hockey, n.d.).[7] More generally, the adoption of venue naming rights in Europe seems to have been more a matter of a culturally "diffuse" learning process – with neoliberalized event cities, towns, and their sporting and entertainment organizations absorbing ideas from each other (and from cities in the rest of the world: see e.g. Crompton & Howard, 2003) – than an example *par excellence* of systematic transurban policy mobility (cf. Klauser, 2011; McCann, 2013). In our data, the only notable exceptions to this observation were the pre- and post-event financial arrangements around the German stadium building boom in the mid-2000s.

That said, in a broad chronological picture the documented national trends, peaks and stagnation phases in the diffusion of venue name sponsorship correlate with the dynamics of sporting and entertainment venue construction. Various dimensions of this linkage are readable in Figure 3, which compares naming patterns in building stock from before and after the advent of venue name sponsorship in Europe. Overall, the considerably higher sponsor-named percentage for post-1990 venues (59.2%) than older facilities (24.5%) reflects the tendency to put revenue gained by name sponsorship deals towards covering the costs of construction. Arguably mirroring a gradual cultural acceptance of venue naming rights, as well as the several facilities built and refurbished for Germany's World Cup tournament, and the pre-crisis economic boom of the new millennium, in 2005–06 alone as much as 8.8% (27 venues) of the entire corpus of venues entered their first naming rights deals. Since the doldrums of the late-2000s, the European growth rate in naming rights deals has been considerably steadier and slower. In itself, the fact that only 10 out of 115 indoor arenas in our corpus are pre-1960 constructions explains the much higher propensity for naming rights in this event facility subcategory. Overall, the chronological logic behind the (non-)adoption of naming rights deals is relatively straightforward: on the one hand, new and state-of-the-art venues attract potential sponsors while, on the other

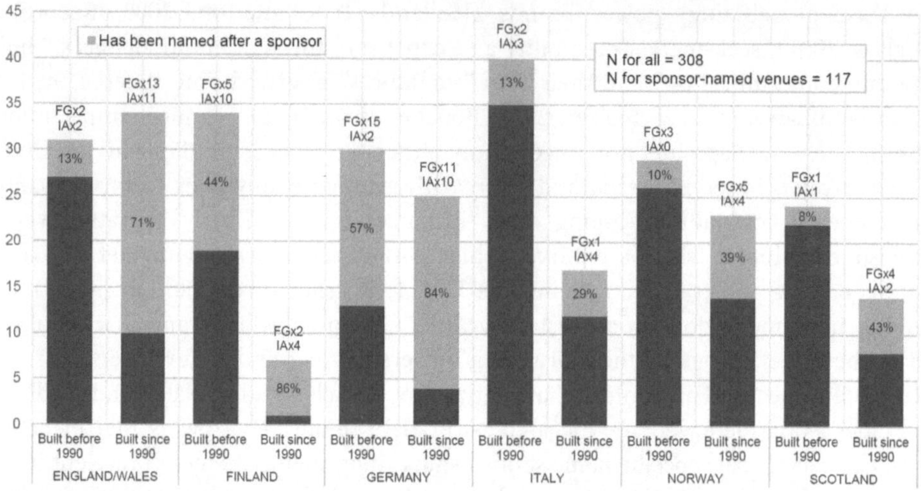

Figure 3. Football grounds and indoor arenas built before and since 1990, including percentages of venues that have been named after one or more corporate sponsors.

hand, the longer the history of a venue is, the less likely it is to have been renamed after the highest corporate bidder.

This brings us to a second set of explanatory factors, touched upon several times above, pertaining to the emergence of variably vociferous fan movements in different countries. Not uncommonly, venue naming rights deals have been targeted in these groups' criticism of the excessive commercialization of football. While venue and team owners almost by definition have the upper hand in (re)naming their facilities, fans, as a stakeholder group, have at least occasionally had a say in decisions related to venue naming. Intriguingly in this regard, this study has indicated considerable differences between the three studied contexts with the most internationally followed football leagues: England and Wales, Italy, and Germany. In the former two contexts, mixed attitudes to the diffusion (or rather non-diffusion in the Italian case) of naming rights have been typical, whereas Germany's situation (with some notable exceptions) is characterized by more market-pragmatist stances. Expectedly, the situation in England has irritated many market-favorable pundits. With James Lawton's (2011) praise of "the Germans [...] who remain the last hope of financial reality [in] Europe," many Newcastle fans' ferocity against the renaming of St James' Park is dismissed as follows: "[T]he idea that the name of a football stadium can ever again be more important than a major injection of income is at the very best quaint" (Lawton, 2011). The fact remains, however, that select symbols steeped in the game's communal historicity have in the last two decades been effectively turned into marketing assets, especially in the context of the English Premier League (Russell, 2006). In the country of origin of the world's most-watched spectator sport, many old stadiums have developed into veritable football pantheons, attracting tourists and sports pilgrims. In the words of the Premier League's executive chairman, Richard Scudamore: "The more global you get, the more local you need to be" (as cited in Ebner, 2013). As long as upholding tradition can be seen to make sense in financial and marketing terms, and the direct commodification of inherited venue names is simultaneously perceived by many as a threat, it is unlikely that the sale of naming rights will develop into the only popular option in naming sporting landscapes.

Conclusions

Increasingly, not only sporting and entertainment facilities, but also multiple other types of urban landscapes are being subjected to naming rights arrangements (see e.g. Rose-Redwood, 2011; see also this issue's introduction). Nonetheless, the status of entertainment and particularly sports venues at the forefront of the entire corporatized landmark language phenomenon seems undeniable. In Europe, sporting facilities have been in the vanguard of implementing naming rights deals since the early-1990s, and particularly in the first decade of the new millennium. It is, however, symptomatic of the relative tardiness in the phenomenon's propagation, especially across the continent's older football grounds, that in all countries in focus protests against venue naming rights have surfaced at least occasionally.[8] Usually, the less historic a venue is, the more (promised) sponsorship money that is set to benefit venue tenants or owners through a name sponsorship deal, and the less (ethically or otherwise) problematic the sponsoring company's engagement is generally seen to be by locals, the easier it is to implement venue name sponsorship without facing insurmountable opposition.

Beyond this general picture, this article has documented a range of national peculiarities related to economic and cultural circumstances in the (non-)diffusion of venue naming rights deals across a European patchwork of states, cities and towns of varying sizes and economic fortunes. In Norway, for instance, the relative health of public finances has apparently meant less financial pressure for the sale of naming rights, whereas in Italy the incidence of less family-friendly phenomena in and around football stadiums has made potential sponsors reluctant to invest in naming rights. In Finland and Germany, meanwhile, the prevalence of market-pragmatic attitudes has paved the way for the adoption of corporatized names. With the venue naming rights trend, the urban effects of neoliberalism have varied considerably between the six contexts studied. Locally embedded interactions between redevelopment projects related to one-off and spatio-temporally diffuse sporting events, country- and locality-specific institutional responses, and varyingly intense civic contestations of this particular culturally inflected aspect of the commodification of space, all illustrate the locally path-dependent dimensions of neoliberal urbanism (see Brenner & Theodore, 2002; Schönig & Schipper, 2016).

To date, very few if any comparative and urban theory-informed analyses have existed on the European-scale spread of the name-sponsoring phenomenon. With our focus on six select contexts and only major sporting and entertainment venues, this study's interpretations should not be seen as exhaustive, but have indicated some significant trends in geographical diffusion. In addition to our Europe-specific empirical findings, we hope to have signaled promising directions for future research agendas and designs in urban place-name scholarship. For urban geographers, much theoretical and empirical interest lies in the documented community-rooted topophilic (Bale, 2000; Tuan, 1974) or indeed toponymophilic (cf. Kostanski, 2009) meanings often set against the hypercommodification of venues, as well as in the functions of inherited toponyms in nostalgia-oriented heritage production and place marketing. Clearly, more research is also needed on the effects of particular regulatory and legislative frameworks on the diffusion of name sponsorships as well as reasons for the counter-trend in which venue naming rights not only proliferate but can also lapse. This research has also touched upon the promises and grievances of spatial naming rights in urban contexts of varying sizes and degrees of connectivity to major hubs and flows of the capitalist world economy. Many of our findings have shown the symbolic hierarchization of space between, on the one hand, affluent urban areas allied with corporate giants and, on the other hand, those left bereft of the advantages of extremely lucrative and more long-standing naming rights deals. One striking feature in many smaller European urban settlements concerns the serial, opportunistic and short-term re-labeling of venues through separate contracts, amounting in Bezold's (2013) terminology to an incapability of utilizing naming rights as a sustainably "strategic" partnership and marketing communication tool. While contemporary metropolises, cities, and towns are characteristically in the same neoliberal boat of aspiring for growth, deploying austerity measures and seeking to plug budget holes, the differing prerequisites for toponymic sponsorship arrangements call for further research into their unequal spatial consequences in Europe and beyond.

Notes

1. The deal was extended in 2012, for a further £150 million, in exchange for five more years of shirt sponsorship, and stadium naming rights until 2028 (Arsenal Football Club, 2012).
2. *Hartwall Areena* is now known as *Hartwall Arena* (with non-Finnish spelling), while *LänsiAuto Areena* later became *Barona Areena* (2009–15) and most recently *Espoo Metro Areena*.
3. In order to safeguard FIFA's own lucrative sponsorship commitments from being violated by "ambush marketing" endeavors, ironically eight name sponsorship deals had to be temporarily rescinded during the 2006 World Cup.
4. The largest Scottish football ground with a sponsored name has a capacity of 8,023 (the *Paisley 2021 Stadium*), which is less than half of the average capacity of our Scottish football ground corpus (16,893).
5. There are 27 double, 14 triple and 3 quadruple name sponsorships per venue in our corpus.
6. These trends go far beyond the scope of our present corpus, with examples such as 梅赛德斯-奔驰文化中心 (the *Mercedes-Benz Arena*) in Shanghai, China, the *Mercedes-Benz Stadium* in Atlanta, Georgia, and the *Mercedes-Benz Superdome* in New Orleans, Louisiana, not to mention the *Allianz Park* in London, England, *Allianz Parque* in São Paulo, Brazil, the *Allianz Riviera* in Nice, France, the *Allianz Stadion* in Vienna, Austria, and the *Allianz Stadium* in Moore Park, Australia.
7. This name sponsorship lasted for roughly one and half years and was replaced by *Data Group Areena*, based on a three-year contract.
8. In the U.S., meanwhile, spatial naming rights have, in most cases, been accepted with much less opposition (see Bezold, 2013; Leeds, Leeds, & Pistolet, 2007). Notable exceptions to this rule include considerable public opposition to the renaming of San Francisco's now demolished *Candlestick Park* after corporate sponsors in 1995–2008, and protests against the corporate name of *Invesco Field at Mile High* in Denver in 2001–11.

Acknowledgments

The authors are grateful to Roosa Wingström, Kristiina Laurikainen, and Sanni Vuorinen for their help while preparing this article.

Disclosure statement

No potential conflict of interest was reported by the authors.

ORCID

Guy Puzey http://orcid.org/0000-0003-2111-5707

References

1. FC Nürnberg spielt künftig im Max-Morlock-Stadion. (2017, May 14). *Kicker Online.* Retrieved from http://www.kicker.de/news/fussball/2bundesliga/startseite/678001/artikel_1-fc-nuernberg-spielt-kuenftig-im-max-morlock-stadion.html

Arsenal Football Club. (2012, November 23). Emirates and Arsenal agree new £150m deal. Retrieved from http://www.arsenal.com/Emiratesnewdeal

Arsenal name new ground. (2004, October 5). BBC sport. Retrieved from http://news.bbc.co.uk/sport1/hi/football/teams/a/arsenal/3715678.stm

Bale, John. (2000). The changing face of football: Stadiums and communities. *Soccer & Society, 1* (1), 91–101.

Bezold, Thomas. (2013). Naming rights for sports stadiums: Sponsor versus club—Who makes the proper brand. In Stephan Sonnenburg & Laura Baker (Eds.), *Branded spaces: Experience, enactments and entanglements* (pp. 121–131). Wiesbaden: Springer.

Bielzer, Louise. (2013). Corporate branded spaces: A glance at today's diversity and some historic origins. In Stephan Sonnenburg & Laura Baker (Eds.), *Branded spaces: Experience, enactments and entanglements* (pp. 89–108). Wiesbaden: Springer.

Bouchet, Patrik, Hillairet, Dieter, & Bodet, Guillaume. (2013). *Sport brands.* Abingdon: Routledge.

Brenner, Neil, & Theodore, Nik. (2002). Cities and the geographies of "actually existing neoliberalism". *Antipode, 34*(3), 349–379.

Broudehoux, Anne-Marie. (2017). *Mega-events and urban image construction: Beijing and Rio de Janeiro.* Abingdon: Routledge.

Church, Andrew, & Penny, Simon. (2013). Power, space and the new stadium: The example of Arsenal Football Club. *Sport in Society, 16*(6), 819–834.

Cleland, Jamie, & Dixon, Kevin. (2015). "Black and whiters": The relative powerlessness of "active" supporter organization mobility at English Premier League football clubs. *Soccer & Society, 16*(4), 540–554.

Crawford, Garry. (2004). Consuming sport: fans, sport and culture. London: Routledge.

Crompton, John, & Howard, Dennis. (2003). The American experience with facility naming rights: Opportunities for English professional football teams. *Managing Leisure, 8*(4), 212–226.

Demir, Robert, & Söderman, Sten. (2015). Strategic sponsoring in professional sport: A review and conceptualization. *European Sport Management Quarterly, 15*(3), 271–300.

Dickson, Andrew. (2013, January 24). Naming rights deal close. *Rangers Football Club.* Retrieved from https://rangers.co.uk/news/headlines/naming-rights-deal-close/

Ebner, Sarah. (2013, July 2). History and time are key to power of football, says Premier League chief. *The Times.* Retrieved from http://www.thetimes.co.uk/tto/public/ceo-summit/article3804923.ece

Gaffney, Christopher Thomas. (2008). *Temples of the earthbound gods: Stadiums in the cultural landscapes of Rio de Janeiro and Buenos Aires.* Austin, TX: University of Texas Press.

Gerke, Markus. (2017). 'Supporters, not consumers.' Grassroots supporters' culture and sports entertainment in the US. *Sport in Society*, 1–14. Advance online publication.

Gir ny stadion til Aalesund. (2002, July 10). *VG*. Retrieved from http://www.vg.no/sport/fotball/1-divisjon/gir-ny-stadion-til-aalesund/a/5652823/

Giulianotti, Richard. (2002). Supporters, followers, fans and flaneurs: A taxonomy of spectator identities in football. *Journal of Sport and Social Issues*, *26*(1), 25–46.

Giulianotti, Richard. (2011). Sport mega events, urban football carnivals and securitised commodification: The case of the English Premier League. *Urban Studies*, *48*(15), 3293–3310.

Gratton, Chris, Shibli, Simon, & Coleman, Richard. (2005). Sport and economic regeneration in cities. *Urban Studies*, *42*(5–6), 985–999.

Harvey, David. (1989). From managerialism to entrepreneurialism: The transformation in urban governance in late capitalism. *Geografiska Annaler: Series B—Human Geography*, *71*(1), 3–17.

Herstein, Ram, & Berger, Ron. (2013). Much more than sports: Sports events as stimuli for city re-branding. *Journal of Business Strategy*, *34*(2), 38–44.

Hognestad, Hans Kristian. (2015). "Rimi Bowl" and the quest for authenticity: Fan autonomy and commodification in Norwegian football. *Journal of Sport and Social Issues*, *39*(2), 139–154.

Hollands, Robert, & Chatterton, Paul. (2003). Producing nightlife in the new urban entertainment economy: Corporatization, branding and market segmentation. *International Journal of Urban and Regional Research*, *27*(2), 361–385.

Jackson, Kate. (2014, September 3). Rangers investor Mike Ashley bought naming rights of Ibrox Stadium for just £1. *Scottish Daily Record*. Retrieved from http://www.dailyrecord.co.uk/sport/football/prospective-rangers-owner-mike-ashley-4161063

Kalpa Hockey. (n.d.). Tervetuloa trust Kapital Areenalle! Retrieved from http://www.kalpa.fi/2012-2013_uutinen/tervetuloa-trust-kapital-areenalle

Kennedy, Peter, & Kennedy, David. (2010). "It's the little details that make up our identity": Everton supporters and their stadium ballot debate. *Soccer & Society*, *11*(5), 553–572.

Kennedy, Peter, & Kennedy, David. (2012). Football supporters and the commercialisation of football: Comparative responses across Europe. *Soccer & Society*, *13*(3), 327–340.

Klauser, Francisco. (2011). The exemplification of "fan zones": Mediating mechanisms in the reproduction of best practices for security and branding at Euro 2008. *Urban Studies*, *48*(15), 3203–3219.

Klingmann, Anna. (2007). *Brandscapes*. Cambridge, MA: MIT Press.

Korkala, Sami. (2006, September 24). Liikuntapaikkojen sponsorinimiä on vaikea myydä. *Oulu-Lehti*, p. 3.

Kostanski, Laura. (2009). *"What's in a name?" Place and toponymic attachment, identity and dependence. A case study of the Grampians (Gariwerd) National Park name restoration process* (Unpublished doctoral dissertation). Ballarat: University of Ballarat.

Lauermann, John. (2016). Municipal statecraft: Revisiting the geographies of the entrepreneurial city. *Progress in Human Geography*, 030913251667324.

Lawton, James. (2011, November 11). Newcastle fans: Deal with it. That's modern football for you. *Independent*. Retrieved from http://www.independent.co.uk/sport/football/news-and-comment/james-lawton-newcastle-fans-deal-with-it-thats-modern-football-for-you-6260279.html

Leeds, Eva Marikova, Leeds, Michael A., & Pistolet, Irina. (2007). A stadium by any other name: The value of naming rights. *Journal of Sports Economics*, *8*(6), 581–595.

Light, Duncan, & Young, Craig. (2014). Toponymy as commodity: Exploring the economic dimensions of urban place names. *International Journal of Urban and Regional Research*, *39*(3), 435–450.

Man Utd rule out renaming stadium. (2005, November 24). *BBC Sport*. Retrieved from http://news.bbc.co.uk/sport1/hi/football/teams/m/man_utd/4465486.stm

McCann, Eugene. (2013). Policy boosterism, policy mobilities, and the extrospective city. *Urban Geography*, *34*(1), 5–29.

McLaughlin, Chris. (2014, November 12). Rangers: Mike Ashley increased loan to Rangers by £1m. *BBC Sport*. Retrieved from http://www.bbc.co.uk/sport/football/30016946

Medway, Dominic, & Warnaby, Gary. (2014). What's in a name? Place branding and toponymic commodification. *Environment and Planning A*, *46*(1), 153–167.

Merkel, Udo. (2012). Football fans and clubs in Germany: Conflicts, crises and compromises. *Soccer & Society*, *13*(3), 359–376.

Molineux could be renamed. (2008, August 22). *Express & Star*. Retrieved from https://www.expressandstar.com/sport/2008/08/22/molineux-could-be-renamed

Peck, Jamie, & Theodore, Nik. (2010). Mobilizing policy: Models, methods, and mutations. *Geoforum*, *41*(2), 169–174.

Pradel i Miquel, March, Paalzow, Anders, & Martin-Brelot, Hélène. (2013). Policies for small and large cities. In Sako Musterd & Zoltán Kovács (Eds.), *Place-making and policies for competitive cities* (pp. 173–190). Oxford: Wiley.

Raento, Pauliina, & Douglass, William A. (2001). The naming of gaming. *Names: A Journal of Onomastics*, *49*(1), 1–35.

Rivrud, Kristin, & Lilleeng, Sverre. (2008, September 1). Komplett.no Arena må bytte navn. *NRK*. Retrieved from https://www.nrk.no/vestfold/komplett.no-arena-ma-bytte-navn-1.6200705

Rose-Redwood, Reuben. (2011). Rethinking the agenda of political toponymy. *ACME: An International E-Journal for Critical Geographies*, *10*(1), 34–41.

Russell, Dave. (2006). "We all agree, name the stand after Shankly": Cultures of commemoration in late twentieth-century English football culture. *Sport in History*, *26*(1), 1–25.

Schönig, Barbara, & Schipper, Sebastian (Eds.). (2016). *Urban austerity: Impacts of the global financial crisis on cities in Europe*. Berlin: Theater der Zeit.

Smith, Andrew. (2016). *Events in the city: Using public spaces as event venues*. New York: Routledge.

SponsorPitch. (2011). The six golden rules of naming rights. Retrieved from https://sponsorpitch.com/posts/the-six-golden-rules-of-naming-rights

Tuan, Yi-Fu. (1974). *Topophilia: A study of environmental perception, attitudes, and values*. Englewood Cliffs, NJ: Prentice-Hall.

Viljamaa-Laakso, Marja. (1999). Markkinavoimat, virtuaalitodellisuus ja paikannimet. In Sirkka Paikkala, Ritva Liisa Pitkänen, Peter Slotte, & Eeva-Liisa Stenhammar (Eds.), *Yhteinen nimiympäristömme: Nimistönsuunnittelun opas* (pp. 82–87). Helsinki: Kotimaisten kielten tutkimuskeskus.

Vuolteenaho, Jani, & Ainiala, Terhi. (2009). Planning and revamping urban toponymy: Ideological alterations in the linguistic landscaping of Vuosaari suburb, eastern Helsinki. In Lawrence D. Berg & Jani Vuolteenaho (Eds.), *Critical toponymies: The contested politics of place naming* (pp. 227–251). Farnham: Ashgate.

Vuolteenaho, Jani, & Kolamo, Sami. (2012). Textually produced landscape spectacles? A Debordian reading of Finnish namescapes and English soccerscapes. *COLLeGIUM: Studies across Disciplines in the Humanities and Social Sciences*, *13*, 132–158.

Walsh, Adrian, & Giulianotti, Richard. (2007). *Ethics, money and sport: The sporting Mammon*. Abingdon: Routledge.

Woisetschläger, David M., Haselhoff, Vanessa J., & Backhaus, Christof. (2014). Fans' resistance to naming right sponsorships: Why stadium names remain the same for fans. *European Journal of Marketing*, *48*(7/8), 1487–1510.

Yurchak, Alexei. (2000). Privatize your name: Symbolic work in a post-Soviet linguistic market. *Journal of Sociolinguistics*, *4*(3), 406–434.

Scalar tensions in urban toponymic inscription: the corporate (re)naming of football stadia

Dominic Medway, Gary Warnaby, Leah Gillooly and Steve Millington

ABSTRACT

This article examines issues of scale in urban toponymic inscription. The specific focus of inquiry is toponymic commodification, whereby corporate brand names of international scope are imposed on English football stadia and their locally embedded fan communities. We employ primary data relating to three football clubs in the Greater Manchester conurbation, all of which have sold their stadium naming rights to corporate entities. Drawing on fans' perspectives, our findings initially surface the scalar tensions arising from such occurrences. We explore how football club authorities attempt to manage these tensions; first through efforts to embed corporate names into the fabric of urban communities, and second by using commemoration to valorize notions of the "local" for their fan base. The article concludes by discussing how our findings deepen understanding of critical toponymies, particularly in terms of theorizing scale and shedding light on the workings of neoliberal agendas for controlling urban space.

Introduction

Since the early 1990s, there has been growing academic interest in critical toponymy, which lies at the intersection of place naming, place making and power (Vuolteenaho & Berg, 2009). Much of this work considers the political dimensions that underpin the urban toponymic landscape, particularly via street naming (e.g. Azaryahu, 1996; Karimi, 2016; Rose-Redwood, 2008a, 2008b; Shoval, 2013), and as a means of power, identity and nation-building by government institutions in postcolonial or post-authoritarian/totalitarian contexts (e.g. Duminy, 2014; Light & Young, 2014; Yeoh, 2009). However, as Hagen (2011) notes, "[a]lthough often implicit in toponymic studies, theorizations of scale are seldom engaged in a sustained manner" (p. 24). This is manifest in a paucity of research into how constructions of scale are deployed in urban toponymic practice by relevant place stakeholders.

Our paper addresses this lacuna by investigating the inter-scalar power relationships involved in the (re)naming of football stadia by sponsoring corporate entities – or what Boyd (2000) refers to as "selling home." The spatial remit of these sponsoring brands can often be far wider (typically international or global) than that of the clubs concerned, which have been described as "small businesses that are locally embedded" (Edensor & Millington, 2008, p. 188). In particular, we focus on scalar tensions arising from the juxtaposition of internationally oriented brands as stadium toponyms with the typically strong connections of football clubs to urban locales and their populations. This emphasizes how the construction of sporting identity continues to resist reductionist accounts by clinging onto spatially rooted distinctiveness, usually within urban contexts (Edensor & Millington, 2008; Williams, 2006). Here, we can turn to Andrews and Ritzer's (2007) critique of the "constitutive inter-relationship between the global and the local" (p. 137), which avoids flattening accounts of globalization's dominance by acknowledging local agency.

Counter to the above arguments, we should not romanticize the local as a site of resistance, as this deflects from the global and "imperialistic" (Andrews & Ritzer, 2007) ambition(s) of some locally embedded institutions and organizations, not least those football clubs that consistently occupy the highest echelons of Europe's various elite leagues (e.g. the English Premier League, the German Bundesliga, and Spain's La Liga). Thus, it has been argued that clubs such as Real Madrid, Bayern Munich, Barcelona, Manchester United and Chelsea, to name but a few, have established themselves as global brands, with legions of fans far removed from the club's locality, and who can be positioned as consumers of its successes and failures (Bodet & Chanavat, 2010). Furthermore, an ongoing and intense bidding war for the global television rights to elite league football has fueled its internationalized marketization (Millward, 2013). Andrews and Ritzer (2007) indicate that in a sporting context such developments are best captured by the deployment of the term "grobal" – an apparent neologism combining the words "growth" and "global", which emphasizes "the fact that there are global processes that overwhelm the local rather than neatly integrating the two" (p. 137).

Building on the above issues, our paper is also concerned with an aspect of critical toponymy deemed worthy of further research (Rose-Redwood, Alderman, & Azaryahu, 2010); namely, toponymic commodification and the selling of naming rights to places. This connects to wider debates regarding the commercialization of urban space, particularly when cities embrace an entrepreneurial agenda (Harvey, 1989). While Light and Young (2015) identify private sector involvement in the naming of sports stadia as a well-established manifestation of such practice, they also recognize that there has been "limited scrutiny of this phenomenon" (p. 439 – for notable exceptions see Boyd, 2000; Church & Penny, 2013; Vuolteenaho & Kolamo, 2012). Thus, by examining the corporate (re)naming of three football stadia within the Greater Manchester conurbation of northwest England, we advance understanding of urban toponymic commodification, and specifically reveal how scalar processes and interactions play a crucial role in this. In doing so, we contribute empirical evidence towards emerging debates regarding the use of corporate toponyms, which have been largely dependent on secondary sources (e.g. Light & Young, 2015; Medway & Warnaby, 2014). Linking back to Harvey's (1989) notion of an urban entrepreneurialism, we also shed light on the machinations of the neoliberal city by demonstrating how institutional elites enact management processes to facilitate the influence of corporate toponyms over urban space.

We begin with a review of literature relating to toponymic commodification, scalar perspectives in critical toponymy, and the local embeddedness of football. Following a brief overview of the three clubs studied, and the data collection/analysis procedures undertaken, our findings are reported under two broad thematic areas: 1) fan perspectives on the scalar tensions arising from corporate (re)naming of their club's stadium; and 2) organizational management of stadia toponyms from the perspective of club authorities. The discussion and conclusion draws the threads of our argument together to shed light on the scalar complexity of urban toponymic inscription in a footballing context.

Toponymic commodification, scale, and football's embeddedness

Previous work has shown that corporate entities can rename areas of urban space for attempted economic gain. Medway and Warnaby (2014) identify how property developers and estate agents in London refer to an area of Kings Cross as the "Regent Quarter" in an attempt to boost property values and attract inward investment. Similarly, Vuolteenaho and Ainiala (2009) demonstrate that the toponymic landscape of Vuosaari, a residential district of Helsinki, reveals "a burgeoning neoliberal tendency in suburban linguistic namescaping, ultimately intended to enthral people as consumers" (p. 247).

We are concerned with corporate toponyms as applied to football stadia. Whilst naming sporting venues after wealthy team owners dates back to the early twentieth century (Leeds, Leeds, & Pistolet, 2007), the purposeful sale of naming rights to stadia by football clubs and their owners is a more recent development. What makes this different from other attempts to rename urban space is that it typically involves the *direct* use of a corporate name, arguably moving into the realm of toponymic *branding*. To illustrate, 14 out of 18 German Bundesliga football stadia had been renamed after corporate brands by 2013 (Woisetschläger, Haselhoff, & Backhaus, 2014); and in 2017, seven of 20 English Premier League stadia also bore a corporate name (Football Tripper, 2017).

However, attempts to apply corporate names to stadia are not always well received by fans. Such a reaction was evident in the 2011 debacle over the renaming of Newcastle United's St James Park ground, so called since 1880, to the "Sports Direct Arena." Significant in this case was the fact that the owner of Newcastle United Football Club, Mike Ashley, also owned the company Sports Direct International plc, with which the stadium naming rights were associated. It was argued by the club that the name change to the Sports Direct Arena signaled the financial potential of Newcastle United to other prospective sponsors, and that, in this respect, the St James' Park name was no longer "commercially attractive" (BBC, 2011). In response, fans took direct action, including painting the original name back onto the perimeter wall of the stadium from where it had been removed (BBC, 2012). As a result, by 2013 the St James' Park name was reinstated (Edwards, 2012).

The corporate renaming of a football stadium arguably moves beyond a straightforward business transaction embedded within conventional sponsorship dealings, when it extends the political and economic influence of private finance into a formerly public realm. This is the recognition that it is not just football stadia themselves that may carry a corporate name, but also the urban space and infrastructure surrounding them, which can become similarly rebranded. Thus, the area around Manchester City's Etihad Stadium is now rarely referred to in Manchester City Council documents and plans by its former name of "Bradford," but as the "Etihad Campus" (see, for example, Manchester City Council Executive, 2017, p. 18). This is reflected in local residents' concerns that the "Bradford" name will be erased with the ongoing march of urban regeneration (Glendinning, 2013). And, as we discuss below, the "Etihad" name has also found its way onto a tram stop of the city's public transport network. These practices are not simply confined to particular sports or countries; in the US, for example, the Dallas Cowboys American football team play at the AT&T Stadium, which is found on AT&T Way.

Such developments appear to demonstrate a "stealthy" (Brown, 2015) approach to the neoliberalization of urban space (Peck & Tickell, 2002), which "is caught up in specific forms of naming that symbolically and materially solidify current (and historical) processes of capitalist accumulation by dispossession" (Berg, 2011, p. 13). In the case of elite international football clubs, these toponymic practices are often ably abetted by super-rich owners, who may seek corporate value, as well as personal status, via financial involvement in their club (Franck, 2010) – indeed, in this sense, the media and supporters' associations often portray football clubs as "playthings" for the rich (Hayward, 2017). Current practice also suggests that corporate value is viewed by a club's ownership as being further enhanced through the imposition of corporate stadium names. This appears especially the case when these stadium names relate to companies in which a club owner may also have financial involvement, as seen with Newcastle United (outlined above) and Manchester City (explained in more detail below).

In another context, Light and Young (2015) identify examples of corporate toponyms in Hungary, manifest in the long-established practice of naming streets after commercial enterprises; a phenomenon more common following the end of state socialism in 1989. Their work also emphasizes the importance of scalar issues, in that some streets with corporate names reflect the embeddedness of *local* companies. However, as Light and Young (2015) note:

> [a] further set of street names reflect the significance of international investment for the Hungarian economy, and the naming practice is clearly linked with promoting Hungary as firmly embedded in global capitalist economic networks (p. 442).

This is evident in toponyms such as Mercedes Street, Nokia Street, and Samsung Square; although demonstrating local as well as global ties, these streets are close to production facilities operated by the companies concerned. Furthermore, Light and Young (2015) indicate the inter-scalar implications inherent in the complex stakeholder relationships underpinning such urban toponymic practice:

> [i]n some cases the companies involved built new roads, which they named after themselves, while in other instances businesses paid local authorities for the naming rights. There are also examples of urban administrations choosing to name streets after companies without payment, perhaps as a means of establishing good relations with them, and in order to embed them (and the investment and employment that they bring) into the local economy (p. 443).

This clearly reveals the importance of scalar considerations in toponymic commodification. However, as noted above, Hagen (2011, p. 23) states that, aside from a few studies (e.g. Alderman, 2003), toponymic research "often ignores how constructions of 'scale' are deployed."

When considering issues relating to scale in wider geographical research, Hagen (2011, p. 24) suggests general agreement "that scale can be conceived as a practice or process rather than an ontological given," thereby connecting to notions of relationality and plurality in scalar matters. Similarly, Brenner (2001) contends that the "politics of scale" should be considered, not in terms of *singular* connotations (i.e. as occurring *within* a relatively bounded geographical area), but from a *plural* perspective, denoting "the production, reconfiguration or contestation of particular differentiations, orderings and hierarchies *among* geographical scales" (p. 600, emphasis in original). This is articulated in terms of "scalar structuration," whereby the "meaning, function, history and dynamics of any one geographical scale can only be grasped relationally, in terms of upwards, downwards, and sidewards links to other geographical scales" (Brenner, 2001, p. 605).

A complementary viewpoint is articulated by Marston, Jones, and Woodward (2005, p. 419), again challenging the "rigidities" of traditional approaches to the concept of scale, arguing for *the local* as an "entry point to understanding 'broader' processes, effectively, examining scale from underneath." This perspective is manifest in the work of Howitt (1993) and Cox (1998), with the latter distinguishing between what he terms *spaces of dependence* and *engagement*. Here, spaces of dependence are defined by more-or-less localized social relations, which define place-specific conditions for material well-being and sense of significance – for our purposes this might include fan affiliation to the local football team. By contrast, spaces of engagement constitute broader sets of relationships of a more global character, which may act to undermine or dissolve spaces of dependence. Drawing on such ideas, we highlight the interactions arising from the imposition of toponyms as corporate brands of international scope on the unique local embeddedness and rooted place identity of football stadia, clubs and their associated fan communities.

In unpacking football's embeddedness, particularly for English clubs with their links to individual cities, towns, and specific urban districts, Edensor and Millington (2008) note the strong ties between teams and the communities in which they were established and have subsequently developed (see also Stone, 2007). These ties are sometimes reflected in the persistence of team nicknames linked to a historic industrial heritage of the town or city in question. Thus, Sheffield United are known as "the Blades," and Stoke City "the Potters" – a reference to the former global dominance of these cities in cutlery and pottery production respectively.

Traditionally, fans have supported their local team, and their cohesion as a community is founded on shared experience and memories (Hand, 2001). Accordingly, football stadia and their environs are rendered as "memory places" (Boyd, 2000) and sites of pilgrimage (Bale, 2000; Church & Penny, 2013), within which a team's triumphs (and disasters) have occurred. Moreover, football stadia are often closely linked to their surrounding areas via road names (Bale, 1989), or other similar references in the local signscape, including the names of small businesses. Local embeddedness is also reflected in place-specific match-day rituals and routines centered on stadia (Edensor & Millington, 2010).

Nevertheless, Edensor and Millington (2008) also recognize that local ties can be eroded by the contemporary restructuring of the football "industry." This takes many forms, not least the growing sponsorship of clubs (especially in the game's higher echelons) by international corporations and their brands, and the associated sale of stadia naming rights, which arguably "dilutes" a club's "local inscription" (Edensor & Millington, 2008, p. 178). Conversely, the influence of football's restructuring forces might sometimes be overstated. For example, the high-profile club Manchester United has ruled out any corporate renaming of its Old Trafford ground, supposedly to preserve the tradition and associations with which this original name is imbued (Ogden, 2015). A more critical reading might contend that clubs like Manchester United have evolved into significant corporate entities in their own right, thereby viewing their stadium name as a lucrative brand asset requiring protection.

Context and method

As noted above, this study focuses on three clubs in the Greater Manchester conurbation, which have all changed their original stadium names to ones associated with a corporate sponsor. The clubs are Manchester City, which, at the time of writing, plays in the English Premier League (the top tier of English professional football); Championship (second tier) team Bolton Wanderers; and Oldham Athletic from League One (third tier).

From 1923–2003, Manchester City played at the club-owned Maine Road stadium to the south of the city center, before moving to a stadium in the east of the city (Manchester City, 2017), originally built for athletics events at the 2002 Commonwealth Games. This new venue was initially named the City of Manchester Stadium, but in 2011 the club negotiated with the City Council (the stadium's owners) for control of the naming rights in return for an increase in rent payments. This paved the way for a 10-year deal with international airline operator Etihad (Taylor, 2011), owned, like the club, by Abu Dhabi's ruling Al Nahyan family. As indicated above, the development of the area around the Etihad Stadium has also been funded by the company and named the Etihad Campus.

Bolton Wanderers played at Burnden Park, a club-owned ground close to Bolton town center, from 1895–1997. The club then moved to the newly-built Reebok Stadium, which it also owned, on the outskirts of Bolton Metropolitan Borough. This venue was initially named after the global brand Reebok, but was subsequently renamed Macron Stadium in 2014, after an Italian-owned international sportswear company (Bonnar, 2014). Oldham Athletic have played at the club-owned[1] Boundary Park stadium since 1899 (The Beautiful History, 2017). In 2014, this venue was renamed SportsDirect.com Park, following a five-year naming rights deal (BBC, 2014) with the UK-based international sportswear manufacturer and retailer Sports Direct International plc.[2] This company is also known for its present and past financial involvement with other UK football clubs, including Newcastle United (see above).

Our findings are drawn from three key data sources collected by two of the authors. First, the environs of the three stadia studied were walked to get a "feel" for the urban areas in which they were situated and observe any toponymic congruity between current and former stadia names, and those of streets and buildings in the vicinity. These data were recorded using field notes and photography. Second, group discussions were undertaken with fans of the three clubs, comprising broad-based topic prompts focused on issues relating to corporate stadium (re)naming. These discussions took place in mutually agreed venues. Participants were recruited via supporters' association representatives and through online sources such as fan message boards, where we gained permission from message board moderators to approach fans. Reflecting a recognized male gender bias among English football fans (EFL, 2015), participants were predominantly male (14 males and 4 females). We also attempted to gain the views of adult fans across a wide variety of age groups, with participants ranging between 20 and 69 years in age. Third, to obtain an organizational perspective, interviews, each lasting approximately one hour, were conducted with senior executives from the three clubs (one at Bolton, one at Oldham, and two together at Manchester City.)[3] Both group discussions and interviews were recorded and transcribed.

Our epistemological approach to the collection and analysis of data accords with an interpretivist philosophy, emphasizing the importance of how language, discourse and meaning are socially constructed (Myers, 2013). Consistent with Abdallah and Langley (2014) and Heracleous (2006), we began from the position of viewing all data (i.e. group discussion/executive interview transcripts and field notes) as text. Each author independently reviewed the data, looking for relevant points of interest. Following the tenets of inter-coder reliability (Campbell, Quincy, Osserman, & Pedersen, 2013), the authors then collectively negotiated their interpretations of the data into the aggregated key themes explored below.

Scalar tensions and corporate (re)naming: fan perspectives

All three clubs studied were open to inter-scalar toponymic tensions (Hagen, 2011; Light & Young, 2015), arising from brand names with international presence and scope being applied to their stadia – and in the case of Bolton, more than once. Such brand impositions on stadium spaces hold the potential to *conflict* with locally embedded fan communities, for whom the stadium is a nexus of regular interaction (see, for example, Edensor & Millington, 2010). "Conflict," in this sense, equates closely to ideas in scalar geography outlined above, relating to how spaces of engagement can undermine and dissolve spaces of dependence (Cox, 1998). However, fan comments in relation to stadia renaming suggested a distinction between the perspectives of Oldham fans tied to a long-established stadium, and those of Bolton and Manchester City, whose stadia had both been newly-built and involved relocations.

Corporate toponyms and the long-established stadium

From the outset, many Oldham fans discussed a perceived irrelevance of the SportsDirect.com Park toponym:

> Boundary Park *is* Oldham Athletic. SportsDirect.com can be any ground in the country. They can go and sponsor anybody else if they want to. They're not tied to us at all... there isn't another Boundary Park (Oldham fan a).

Here, the imposition of a corporate toponym highlights the scalar discrepancy between an international brand and local community interests. Tension surfaces in the perceived potential of the corporate name to render Oldham Athletic's stadium as a space of "nowhereness" (Arefi, 1999), or "any ground in the country." As one fan explained:

> Boundary Park; it's got something about it. It's more a sports name. It's a park. The area is part of the town. Sports Direct is not part of the town. (Oldham fan b).

Similarly, another fan identified how the Boundary Park name "defines a geographical area that people can relate to, whereas you can't do that with a company" (Oldham fan c).

Corporate names were therefore regarded as little more than temporary, removable "place holders" that could not dislodge the original toponym:

> Boundary Park... it wouldn't be SportsDirect.com Park because it's not a permanent name. Boundary Park's a permanent name, if you know what I mean (Oldham fan a).

This dislike of – and unwillingness to engage with – the corporate stadium name encapsulated fans' protection of, and pride in, their locally embedded and rooted identity (Edensor & Millington, 2008; Stone, 2007):

> Oldham fan d: "Some people are more ferocious about their area and their name than other areas."
>
> Oldham fan e: "It's our pride isn't it."
>
> Oldham fan d: "It's like Stoke fans are really ferocious, Newcastle fans are really ferocious."
>
> Oldham fan b: "They need their name, they need their identity."
>
> Oldham fan e: "It's because those clubs are more rooted in the local environment. If you look at Man United, there's more fans outside Manchester than there are in."
>
> Oldham fan d: "They've got history, but they haven't got local support. [From] Salford maybe, but otherwise they're coming from all over the world."

It can be seen here that Oldham fans also contrasted their situation with that of bigger clubs like Manchester United. Whilst it is clearly unrealistic to suggest that Manchester United has not got "local support," the point remains that such support is inevitably rendered less significant when set against a backdrop of grobal forces (Edensor & Millington, 2008; after Andrews & Ritzer, 2007). These forces are geared to promoting and monetizing the football club as a global brand (Hamil, 2008; Lee, 1998), which can result in an international fan base that outnumbers and overshadows that drawn from the local area.

Corporate toponyms and newly-built stadia

For Manchester City and Bolton fans, scalar tensions were less defined, perhaps reflecting the fact that both the Etihad and Macron stadia are recently built, and have involved relocations from the clubs' former "homes." Consequently, the corporate name was less likely to be in direct conflict with any deep, topophilic bond (Tuan, 1974) between fans and their new stadium and its surroundings, as this had not had time to fully develop. This accords with Edensor's (2015, p. 83) suggestion that the emotional and affective sense of belonging that typifies the connectedness of fans and football stadia is embedded through "repetitive practical, embodied engagement," and therefore takes time to "settle" and "knit" in new locations. As one Bolton fan stated:

> We haven't got the history yet [at] the Reebok/Macron. Maybe in 50 years' time there might be, because the history will be there; but because it's a new stadium, or still quite a new stadium, it's not got the history (Bolton fan a).

Despite this, the same fan contrasted the local connections of the former Reebok corporate stadium name (a global brand, but originating in Bolton), with the absence of a similar local heritage for Macron:

> Reebok was kind of a Bolton firm..., it was at least keeping it within Bolton. It wasn't like it was branded or something, like it is now (Bolton fan a).

Another fan noted that there had been a loss of local connectivity and embeddedness for the club through this change of corporate toponym, claiming, "We've lost that local side of it to an Italian company" (Bolton fan b).

Stadium newness also deflected concerns over the Etihad name amongst Manchester City fans:

> For us, it was slightly different because it's a new stadium, we had no history (Manchester City fan a).

In addition, fans recognize the fact that since 2008 the Etihad name can be associated with significant financial investment in the club by its current owners, the Al Nahyan family. This has clearly contributed to the team's most recent success in the form of FA and League Cup wins in 2011 and 2016 respectively, and two Premier League titles in 2012 and 2014:

> ...because that's a new stadium, I didn't really give a monkeys, if it's more money to buy players... So changing it to something like the Etihad, I think we're okay with: a) because it's new, b) because it's a direct relationship between the owners of the club, and not many people were unhappy with the owners (Manchester City fan b).

Notwithstanding Manchester City fans' general acceptance of the Etihad stadium name, the brand's international scope has inevitably pushed the club further onto a global stage, to rival the media profile of footballing neighbors Manchester United. This has attracted a new breed of football flâneurs (Giulianotti, 2002) and "tourist" fans (Linden & Linden, 2017, p. 159) that are not always welcomed by existing supporters:

> We're getting tourists now... There is an uneasy relationship between those who have been there a long time and those who are new to it. I feel that tension sometimes (Manchester City fan c).

There is also an apparent "othering" discourse at play here, whereby genuine fan identity is constructed as something time-served and locally rooted, compared to new fans from outside the area who are devoid of such credentials.

Prior to Etihad's involvement in Manchester City, Edensor and Millington (2008) optimistically proposed that the club was attempting to apply a "glocal" strategy of "nuanced embeddedness" to its business, combining a global focus with continued sensitivity to local fan bases. However, fans' views above suggest that ongoing tensions between global and local fan bases, catalyzed in part by the introduction of the Etihad toponym, are becoming increasingly challenging to manage effectively without a "flattening out of place, identity and culture" (Edensor & Millington, 2008, p. 173) at the local level, and an inevitable drift of the club towards grobalization.

Imagined resistance to corporate stadium toponyms

The above discussion indicates that a club's move to a new stadium denudes the local embeddedness of its fan base to some degree. Others suggest this is manifest in the disruption of fans' match day routines and associated geographies of practice in new stadia settings (Edensor & Millington, 2010). Such a weakening of place connectedness appears to temper fans' hostile responses to the renaming of their *new* stadium after an international brand, as locally embedded notions of identity have not had time to develop, and are not, therefore, perceived as under threat.

Conversely, when imagining the hypothetical scenario of their former stadia assuming a corporate brand name, Bolton and Manchester City fans alike indicated they would view this as a more direct threat to a historically accumulated sense of local embeddedness and spatially connected identity. This resonates with ideas regarding the importance of an individual's "interactional past" with a physical site in the development of place attachment to it (Milligan, 1998). As two Bolton fans explained:

> Bolton fan c: "I would have been totally disgusted. I don't think I could have accepted that..."
>
> Bolton fan a: "It's the history that comes with that. Like Burnden Park was, you know, years, 100 years of history that is there at that stadium, that's always been Burnden."
>
> Bolton fan c: "That's Burnden and that's where we are, who we are."
>
> Bolton fan a: "Yeah, exactly, that's like your identity embedded in the stadium."

Manchester City fans were equally certain in their imagined resistance to any corporate renaming of the former Maine Road stadium, where the team had played for 80 years, emphasizing how this would have disrupted a temporally and spatially accumulated sense of connection with the club:

> Yeah, I would have been a lot less happy if it had been Maine Road... because one of the things that makes football such, when we have this connection to it, is we can go back with those years and those links and so forth. And I think if you'd start interfering with that it weakens the whole relationship you've got with the club (Manchester City fan c).

These narratives of imagined resistance arguably act as a form of psychological "repair and maintenance" work (Graham & Thrift, 2007), allowing fans to articulate a tough rhetoric of toponymic resistance that preserves a stability between signs and spatial order in a hypothetical sense. This presents them as loyal supporters with a sense of club heritage. At the same time, the imagined nature of such toponymic resistance allows room for fan discourses to acknowledge that, in reality, the corporate name of their new stadium (and any associated investment) is part of the modern game. Furthermore, it could disadvantage the club if these opportunities were rejected:

> I mean it's a sign obviously of moving towards more modern times, isn't it? It is a branded stadium. Obviously, that's more money… Burnden Park was Burnden Park because of where it was. And I suppose there's a lot of nostalgia around Burnden… I think it helps keep that era and obviously the memories alive there, but it's a brand new stadium, modern, and it was kind of just keeping up with modernization of sponsorship of the stadium, which obviously other clubs are doing as well (Bolton fan a).

Oldham Athletic have played at the same stadium for over a century, and as evidenced above, their fans articulated a "ferocious" protection and preservation of the original Boundary Park toponym in the face of its corporate replacement. This is perhaps the best evidence that the imagined resistance of Bolton and Manchester City fans to the renaming of their former Burnden Park and Maine Road stadia is not just idle talk.

Organizational management of corporate toponyms

It is also instructive to consider the corporate (re)naming of football stadia from the perspective of club authorities. The critical toponymy literature highlights naming practices as manifestations of political power play, and hegemonic control by institutional authorities within urban space (e.g. Alderman, 2003; Rose-Redwood et al., 2010). Such practices may also face active citizen (or in our specific context, fan) resistance in the form of deliberate, counter-performative toponymic utterances (Rose-Redwood, 2008b), or a more passive inertia and habit through the continued use of historical toponyms by the local populace (Light & Young, 2014).

Previous work on institutional efforts to inscribe space with new toponyms has been largely based on archival work (e.g. Duminy, 2014; Rose-Redwood, 2008b; Shoval, 2013). Here, by contrast, the renaming of stadia was so recent and so fresh in the memory that club executives could be questioned directly on the matter. They were candid on the intent of such initiatives, emphasizing the importance of sponsorship and commercial investment (Buhler, 2006; Emery & Weed, 2006), and that the views of fans were of limited consequence in determining the outcome of corporate (re)naming processes.

One executive (anonymized) articulated his club's approach quite simply, noting, "It's whatever the deal the commercial team is able to strike that will come out on top." Another advocated a slightly more flexible approach:

> I mean ultimately, because we're obviously a private business, the decision was made by the Board at the time. But historically, we do have several focus groups with fans, just to get a flavor and stuff like that. But ultimately the final decision will boil down to the chairman. (Club executive – anonymized).

This suggests that fans had been consulted. In this instance, however, it is implied that this amounted to a token gesture by the club to show that is was acting in a consultative manner, rather than out of a genuine desire to capture fans' views.

Embedding corporate names at the local level

Despite this rather bullish approach, club authorities were clearly aware of the potential for scalar tensions between corporate brands of international scope being used as stadium names, and the local connectedness of fan communities. They were, therefore, keen to highlight attempts to embed those brand names more effectively at a local level. At Manchester City, this has been facilitated through a purposeful strategy of widening the scope of inscription for the Etihad brand name into the regeneration of space and buildings around the stadium, and even a nearby tram stop (Figure 1):

> It wasn't a standalone move… it became the Etihad Campus and the [Etihad] tram stop opened. And at the same time it became the Etihad Stadium (Manchester City executive).

As Church and Penny (2013) indicate in their work on Arsenal Football Club, names that extend the reach of stadium institutions into surrounding spaces in this manner may also help to legitimize the process of public space management.

In addition, the Etihad Campus also includes venues for all-year-round community activities and participation, enabling the Etihad name to become further engrained within the fabric of the local area:

> …building facilities on this site [are] kind of rooted more into the local wider community. And it's not just a football ground at weekends and stuff, but a 7/365 destination (Manchester City executive).

This shows how toponymic commodification can be more successful when there is a degree of "scalar sensitivity" in its introduction, embracing a spatial remit which involves integration of the corporate brand with community life, and in which the

Figure 1. Sign denoting the name of the tram stop nearby the Etihad Stadium.
Source: Authors' image.

stadium itself is but one integrated element of a broader toponymic intervention within an urban district.

Such an approach also represents a scalar inversion of the brand-place interrelationships identified by Pike (2011), using the example of the beer brand Newcastle Brown Ale. Here, connections to Newcastle, realized through incorporation of the toponym within the brand name, arguably act as a marketing platform for increasing the international sales of this consumer product, capitalizing on ideas of authenticity and local provenance. Thus, embeddedness within a particular locale is a promotable asset, leveraging inherent and existing connections, rather than something that needs to be nurtured, as is seen with Etihad and Manchester City. These issues aside, efforts to embed the Etihad brand at the local level appear to be appreciated by fans:

> The owners aren't daft. They've picked that area and they can effectively do what they want. [The city of] Manchester's glad of the investment; and quite rightly so... that investment is coming into Manchester... and it's never going to be fast, but it's regenerating the whole area. And that in turn will give a hinterland for City to be pulling in fans. It's not just about renaming the stadium. It's about almost renaming the whole area (Manchester City fan a).

Unlike Etihad, Macron has not made significant investment and toponymic interventions in the landscape around Bolton Wanderers' stadium, partly because the development opportunities for doing this are limited.

Nevertheless, since 2014 the company has made a concerted effort to build relationships with the local fan base through community action and philanthropic intervention. This includes donating kits to local junior football clubs and clothing to the town's homeless community at Christmas, as well as offering opportunities for fans to win Macron-related experiences, such as visits to the company's factory in Italy. These efforts have also been positively received:

> Bolton fan b: "It's things like that that get noticed, isn't it? Makes them look good... it'll make the fans happy that it's not all take, take, take..."
>
> Bolton fan a: "It's getting involved isn't it, with the community..."

At Oldham Athletic, there was no evidence of Sports Direct attempting to build such connections. On the one hand, this is surprising, as findings show that of the three clubs studied, Oldham Athletic was where scalar tensions arising from the imposition of a brand name on a stadium were felt most acutely by fans. Conversely, the lack of local engagement by Sports Direct may be a result, rather than a cause, of negative fan reactions to the incorporation of this brand into their stadium name – reactions arguably influenced by the fact Boundary Park was a long-established toponym for a stadium that has not, crucially, moved location. As outlined above, it may also be significant that Sports Direct received similar adverse fan reaction when it briefly renamed Newcastle United Football Club's St James' Park stadium (another long-standing venue and toponym) the "Sports Direct Arena" in 2011 (BBC, 2012). This might explain why SportsDirect.com branding was barely evident on the exterior of Oldham Athletic's stadium, with the exception of one small sign over the club shop (Figure 2). There were no signs declaring the official "SportsDirect.com Park" toponym;

Figure 2. Sole exterior sign for SportsDirect.com on Oldham Athletic's stadium.
Source: Authors' image.

and, three years into the sponsorship deal, there seemed little impetus from the company to change this:

> ...they are entitled in the contract to repaint the roof... fully entitled to repaint that and rebrand it. They haven't done it yet..., but that's entirely their choice. It's in the contract but it's at their cost... They're still in the process of looking at it (Oldham executive).

When fan reactions are so negative (or indifferent) towards corporate toponymic inscription, it suggests that the key commercial benefit for a naming rights sponsor in these instances lies with the toponym's media presence, rather than its physical imposition upon stadium space and surrounding areas. However, such media usage requires managing, and executives at all three clubs identified that compliance with the corporate stadium name was something they policed within media and press coverage as part of the contractual obligation to their sponsoring brand.

(Re)embedding the "local" through commemoration

The use of commemoration in toponymic inscription has been associated with strategies by political elites "to construct and institutionalize a hegemonic narrative of history" in urban space (Light & Young, 2015, p. 436). In the case of Bolton Wanderers and Manchester City, club authorities were also keen to invoke commemoration of former grounds and toponyms. However, it appeared this was less about providing a particular version of history, and more concerned with giving a focus for fans' memories and nostalgia that could quell the spatially disruptive and locally de-embedding forces of stadium relocations and associated name changes to internationally positioned brands. The Bolton executive explained how this was the rationale for naming a road outside the club's new stadium "Burnden Way" (Figure 3), after the old Burnden Park ground:

> ...if you move somewhere new like here it's a blank canvas..., but there was always a view that we wanted to retain part of the history... And Burnden Way... I think fans would have been consulted... Although we are moving away, we very much want to return to the heritage of the club from that.

Similarly, at Manchester City, club authorities recognize the power of commemoration as a means of (re)embedding fans within the new stadium location by using artifacts from the former Maine Road ground, along with statues, and the naming of roads and stands, that reference former players and managers from the Maine Road era. This echoes Church and Penny's (2013) work, identifying similar commemorative devices around Arsenal Football Club's new Emirates stadium. Significantly, whilst this approach encourages fans within a new stadium to create a "memory place" (Boyd, 2000) that connects them to a former ground, in the case of Manchester City this is attempted without any overt reference to the Maine Road name, which might arguably undermine the credibility of the Etihad brand as a stadium toponym:

> Manchester City executive: "There used to be a mosaic from Maine Road..., which I now think is in the memorial garden. But it used to be above the main entrance to Maine Road. The Colin Bell Stand is named after a legend of that era. You've got Joe Mercer Way, with a statue at the top to again commemorate a manager from the Maine Road era... And then

Figure 3. Burnden Way sign with Macron Stadium in background.
Source: Authors' image.

> you've got the ring, you know inside the bowl you've got all the names of previous captains which all dates back… It doesn't reference Maine Road."
>
> Interviewer: "It mainly references the players?"
>
> Manchester City executive: "But if you're aware of those players, it takes you back…"

For Manchester City, these commemorative interventions also help to further ameliorate potential scalar tensions resulting from the imposition of Etihad's internationally prominent brand name onto a stadium frequented by supporters who see themselves as locally embedded, "real Mancunians" (Edensor & Millington, 2008, p. 172); especially as the club has implemented a degree of consultation in this process:

> I would say [the owners are] pretty respectful of the history of the club and learnt about the history of the club and so forth… they've asked people who know what they're talking about… they really researched it (Manchester City fan c).

At Oldham, commemoration of the former stadium name is perhaps less of an issue, as the stadium itself has not moved. Whilst a road leading to the newly named stadium is called Boundary Park Road, this is not intentional commemoration by the club, but a long-standing relic of the original toponymic inscription.

Discussion and conclusion

Through an examination of three clubs, this article has revealed the pluralistic scalar tensions arising from the corporate toponymic (re)inscription of football stadia, as well as the associated interplay of internationally focused brands and the locally embedded nature of English football. Echoing the work of Hagen (2011), we would argue that this emphasizes how notions of geographical scale in toponymic application cannot simply be reduced to the assigning of place names to fixed and boundaried concepts of space. Rather, scale and scalar interaction in the toponymic landscape needs to be understood in terms of dynamic and ongoing socio-economic and socio-political processes. Our research reveals that such processes are shaped according to complex relational entanglements, which affect how one scale of understanding may interact with another (Brenner, 2001; Cox, 1998; Howitt, 1993; Marston et al., 2005). The evidence presented above also demonstrates how these entanglements have temporal and locational dimensionality, including: the longevity of a stadium within a specific locale; how long a usurped toponym was previously associated with the football club; and whether a stadium has moved to a different site.

For example, Oldham Athletic fans, whose stadium occupied the same location with the same name for over 100 years, are more "ferocious" in their resistance to a new corporate stadium toponym. This is in contrast to Manchester City fans, based at a new stadium (physically removed from the location of its predecessor), which initially had a name assigned from its former life as an athletics venue. Bolton Wanderers fans also appear far less resistant to the imposition of a corporate toponym. Whilst they may have had 20 years to (re)build a sense of local embeddedness around their "new" stadium, the fact that its name has changed in that time appears to engender a degree of indifference to corporate naming and *re*naming. This is perhaps symptomatic of the relentless and ongoing churn of corporate sponsorship deals in football. As one Bolton fan stated:

> It's not as important. It's just corporate isn't it? If it was being named the Nat Lofthouse Stadium,[4] or something like that, then there'd be a lot of ceremony around that, but it's just literally changing hands, changing sponsorship I suppose (Bolton fan a).

The degree of international coverage and scope for those brands involved in the corporate renaming of stadia also has clear scalar implications. Of the three clubs studied, Manchester City's Etihad stadium name is underpinned by the most significant brand in terms of international reach and value. This, in turn, reflects how Manchester City is promoted through the Etihad brand on a global stage. It raises questions, and tensions, regarding how the club can effectively maintain local embeddedness whilst emphasizing a glocal perspective; or whether the globally-scaled ambition and focus of the Etihad brand carries over to the club, rendering its outlook as more grobal.

Other issues are also relevant in this scalar battleground. One relates to the level at which a football team plays, with more successful clubs such as Manchester City attracting sponsors and brand names that possess greater international presence and exposure. Unless dealt with sensitively, the toponymic application of these brands to stadia and their surrounding urban infrastructure may clash with the views of localized fan bases. Equally, the size and diffuseness of the locale to which a club is tied may be important, particularly if smaller, compact settlements deliver a stronger sense of community and local embeddedness. This may further explain the more resistant attitudes of Oldham Athletic fans towards the corporate renaming of their stadium, compared to those of Manchester City; a possible reflection of the fact that the population of the Metropolitan District of Oldham is less than half that of the City of Manchester (City Population, 2017).

A further complexity in scalar understandings of corporate toponymic inscription relates to how the companies (and in this case clubs) deploying brands as toponyms can embed them more effectively at the local level. This could include urban regeneration by the relevant corporate entity and naming sponsor in and around the area of toponymic inscription, along with associated philanthropic interventions in community-based activities. Equally, invoking commemoration of former toponyms may temper fan resistance to corporate replacements. Whilst it is not the task of this paper – or necessarily our wish – to deliver a "how-to guide" for neoliberal urban intervention, it seems, in this context, that there are tactics that organizations can effectively use to facilitate toponymic commodification without destabilizing topophilic notions of "home."

A final point relates to organizational and institutional control over the use of corporate stadium names. We have identified how club authorities attempt to manage this in media contexts. Yet, they have far less jurisdiction over whether fans use official corporate toponyms when referring to *their* stadium in everyday conversation, or engage with other counter performative (Rose-Redwood, 2008b) and habitual (Light & Young, 2014) toponymic colloquialisms and utterances. As the Oldham Athletic executive explained: "Fans do it deliberately. They prefer to call it Boundary Park still. But we obviously can't control that side of things." Equally, Bolton fans reported sometimes using the former Reebok epithet for their stadium, largely out of habit and inertia in speech. Manchester City fans also recounted the occasional use of alternative toponyms for their current stadium, such as CoMStad (an acronym of the pre-Etihad City of Manchester Stadium) and Eastlands (the name of the site on which the Etihad is built). Toponymic corruptions for the Manchester City stadium, such as Middle Eastlands (a reference to the club's Arab owners), can also be found in use on social media.

Clearly, examining how corporate toponyms are performed, counter-performed, or not performed at all in football fan discourse, and how this links with any scalar tensions relating to the application/imposition of corporate stadia names, is one area for future research. Another direction of inquiry might be to examine how scalar relations between corporate toponyms and urban communities are realized in different contexts. It could be that English football, with its typically localized and loyal fan base, maximizes the potential for scalar tensions to arise through corporate naming practices. Alternatively, those tensions revealed in this study may reflect a more universal urban phenomenon arising from the corporate creep of neoliberal intervention in the toponymic landscape of towns and cities.

Notes

1. Boundary Park was owned by Oldham Athletic between 1899 and 1999, when the ground was sold to Oldham Property Partnerships (OPP), a joint venture company involving Oldham Council. This allowed the football club to raise funds and avoid going out of business. In 2005, the club bought back their ownership of Boundary Park from OPP (MEN, 2013).
2. Since this renaming, the company Sports Direct has been involved in media controversy over its employee working conditions and contractual arrangements (e.g. Goodley & Ashby, 2015a, 2015b). These revelations only began to surface whilst the fieldwork for our study was being undertaken, and were not mentioned in detail by our research participants, specifically those from Oldham.
3. For the sake of brevity, we often use the terms "Oldham" and "Bolton" to refer to "Oldham Athletic" and "Bolton Wanderers" respectively. Conversely, we always use the "Manchester City" name in full, as there is another major football team within Manchester.
4. A legendary Bolton player who played for the club from 1946–1960.

Disclosure statement

No potential conflict of interest was reported by the authors.

ORCID

Leah Gillooly http://orcid.org/0000-0002-1320-2803

References

Abdallah, Charazad, & Langley, Ann. (2014). The double edge of ambiguity in strategic planning. *Journal of Management Studies*, *51*(2), 235–264.

Alderman, Derek. (2003). Street names and the scaling of memory: The politics of commemorating Martin Luther King, Jr within the African American community. *Area*, *35*(2), 163–173.

Andrews, David, & Ritzer, George. (2007). The grobal in the sporting glocal. *Global Networks*, *7* (2), 135–153.

Arefi, Mahyar. (1999). Non-place and placelessness as narratives of loss: Rethinking the notion of place. *Journal of Urban Design*, *4*(2), 179–193.

Azaryahu, Maoz. (1996). The power of commemorative street names. *Environment and Planning D: Society and Space, 14*(3), 311–330.
Bale, John. (1989). *Sports geography*. London: Routledge.
Bale, John. (2000). The changing face of football: Stadiums and communities. *Soccer & Society, 1*(1), 91–101.
BBC. (2011). Newcastle rename St James' Park the Sports Direct Arena. Retrieved from http://www.bbc.co.uk/sport/football/15668207
BBC. (2012). St James' Park graffiti: Michael Atkinson admits painting stadium. Retrieved from http://www.bbc.co.uk/news/uk-england-tyne-17077966
BBC. (2014). Oldham athletic: Boundary Park to become SportsDirect.com Park. Retrieved from http://www.bbc.co.uk/sport/football/28144525
The Beautiful History. (2017). Oldham athletic. Retrieved from https://www.citypopulation.de/php/uk-greatermanchester.php
Berg, Lawrence. (2011). Banal naming, neoliberalism, and landscapes of dispossession. *ACME: an International Journal for Critical Geographies, 10*(1), 13–22.
Bodet, Guillaume, & Chanavat, Nicolas. (2010). Building global football brand equity: Lessons from the Chinese market. *Asia Pacific Journal of Marketing and Logistics, 22*(1), 55–66.
Bonnar, Neil (2014, April 24). Bolton Wanderers' Reebok Stadium to be renamed Macron Stadium. Lancashire Telegraph. Retrieved from http://www.lancashiretelegraph.co.uk/news/11169269.UPDATED__Reebok_Stadium_to_be_renamed_Macron_Stadium/
Boyd, Josh. (2000). Selling home: Corporate stadium names and the destruction of commemoration. *Journal of Applied Communication Research, 28*(4), 330–346.
Brenner, Neil. (2001). The limits to scale? Methodological reflections on scalar structuration. *Progress in Human Geography, 25*(4), 591–614.
Brown, Wendy. (2015). *Undoing the demos*. Cambridge, MA: MIT Press.
Buhler, André. (2006). *Professional football sponsorship in the English Premier League and the German Bundesliga* (Unpublished doctoral dissertation). University of Plymouth, Plymouth.
Campbell, John, Quincy, Charles, Osserman, Jordan, & Pedersen, Ove. (2013). Coding in-depth semi-structured interviews: Problems of unitization and intercoder reliability and agreement. *Sociological Methods & Research, 42*(3), 294–320.
Church, Andrew, & Penny, Simon. (2013). Power, space and the new stadium: The example of Arsenal Football Club. *Sport in Society, 16*(6), 819–834.
City Population. (2017). United Kingdom: Greater Manchester. Retrieved from https://www.citypopulation.de/php/uk-greatermanchester.php
Cox, Kevin. (1998). Spaces of dependence, spaces of engagement and the politics of scale, or: Looking for local politics. *Political Geography, 17*(1), 1–23.
Duminy, James. (2014). Street renaming, symbolic capital, and resistance in Durban, South Africa. *Environment and Planning D: Society and Space, 32*(2), 310–328.
Edensor, Tim. (2015). Producing atmospheres at the match: Fan cultures, commercialisation and mood management in English football. *Emotion, Space and Society, 15*(May), 82–89.
Edensor, Tim, & Millington, Steve. (2008). 'This is Our City': Branding football and local embeddedness. *Global Networks, 8*(2), 172–193.
Edensor, Tim, & Millington, Steve. (2010). Going to the match: The transformation of the match-day routine at Manchester City FC. In Sybille Frank & Silke Steets (Eds.), *Stadium worlds: Football, space and the built environment* (pp. 146–162). Abingdon: Routledge.
Edwards, Luke. (2012, October 09). Newcastle United sponsorship deal with Wonga will see St James' Park reinstated as stadium name. The Telegraph. Retrieved from http://www.telegraph.co.uk/sport/football/teams/newcastle-united/9596399/Newcastle-United-sponsorship-deal-with-Wonga-will-see-St-James-Park-reinstated-as-stadium-name.html
EFL. (2015). Infographic: 2015 fan survey. Retrieved from http://www.efl.com/news/article/2015/infographic-2015-fan-survey-2784954.aspx
Emery, Ross, & Weed, Mike. (2006). Fighting for survival? The financial management of football clubs outside the 'top flight' in England. *Managing Leisure, 11*(1), 1–21.

Football Tripper. (2017). Premier league 2016/2017: Premier League stadiums. Retrieved from http://footballtripper.com/europe/england/premier-league/
Franck, Egon. (2010). Private firm, public corporation or member's association governance structures in European football. *International Journal of Sport Finance*, *5*(2), 108–127.
Giulianotti, Richard. (2002). Supporters, followers, fans, and flâneurs: A taxonomy of spectator identities in football. *Journal of Sport and Social Issues*, *26*(1), 25–46.
Glendinning, Amy. (2013, January 12). Why Bradford will live in our memories. Manchester Evening News. Retrieved from http://www.manchestereveningnews.co.uk/news/local-news/why-bradford-will-live-in-our-memories-889068
Goodley, Simon, & Ashby, Jonathan. (2015a, December 09). A day at 'the gulag': What it's like to work at Sports Direct's warehouse. The Guardian. Retrieved from https://www.theguardian.com/business/2015/dec/09/sports-direct-warehouse-work-conditions
Goodley, Simon, & Ashby, Jonathan. (2015b, December 09). Revealed: How Sports Direct effectively pays below minimum wage. The Guardian. Retrieved from https://www.theguardian.com/business/2015/dec/09/how-sports-direct-effectively-pays-below-minimum-wage-pay
Graham, Stephen, & Thrift, Nigel. (2007). Out of order: Understanding repair and maintenance. *Theory, Culture & Society*, *24*(3), 1–25.
Hagen, Joshua. (2011). Theorizing scale in critical place-name studies. *ACME: An International E-Journal for Critical Geographies*, *10*(1), 23–27.
Hamil, Sean. (2008). Manchester United: The commercial development of a global football brand. In Simon Chadwick & Dave Arthur (Eds.), *International cases in the business of sport* (pp. 114–134). Oxford: Elsevier.
Hand, David. (2001). City 'til I die? Recent trends in popular football writing (Review Article). *Soccer & Society*, *2*(1), 99–112.
Harvey, David. (1989). From managerialism to entrepreneurialism: The transformation of urban governance in late capitalism. *Geografiska Annaler. Series B, Human Geography*, *71* (1), 3–17.
Hayward, Paul. (2017, September 04). Campaigners urge government to clamp down on football club ownership. The Telegraph. Retrieved from http://www.telegraph.co.uk/football/2017/09/04/campaigners-urge-government-clamp-football-club-ownership/
Heracleous, Loizos. (2006). *Discourse, interpretation, organization*. Cambridge: Cambridge University Press.
Howitt, Richard. (1993). 'A world in a grain of sand': Towards a reconceptualization of geographical scale. *Australian Geographer*, *24*(1), 33–44.
Karimi, Ali. (2016). Street fights: The commodification of place names in post-Taliban Kabul City. *Annals of the American Association of Geographers*, *106*(3), 738–753.
Lee, Simon. (1998). Grey shirts to grey suits: The political economy of English football in the 1990s. In Adam Brown (Ed.), *Fanatics! Power, identity and fandom in football* (pp. 32–49). London: Routledge.
Leeds, Eva, Leeds, Michael, & Pistolet, Irina. (2007). A stadium by any other name: The value of naming rights. *Journal of Sports Economics*, *8*(6), 581–595.
Light, Duncan, & Young, Craig. (2014). Habit, memory, and the persistence of socialist-era street names in postsocialist Bucharest, Romania. *Annals of the American Association of Geographers*, *104*(3), 668–685.
Light, Duncan, & Young, Craig. (2015). Toponymy as commodity: Exploring the economic dimensions of urban place names. *International Journal of Urban and Regional Research*, *39* (3), 435–450.
Linden, Henrik, & Linden, Sara. (2017). *Fans and fan cultures: Tourism, consumerism and social media*. London: Palgrave Macmillan.
Manchester City. (2017). Timeline. Retrieved from https://www.mancity.com/fans-and-community/club/club-history
Manchester City Council Executive. (2017, March 8). The eastlands regeneration framework. Report to economy scrutiny committee. Retrieved from http://www.manchester.gov.uk/meetings/meeting/2626/executive

Marston, Sallie, Jones, John, III, & Woodward, Keith. (2005). Human geography without scale. *Transactions of the Institute of British Geographers*, *30*(4), 416–432.
Medway, Dominic, & Warnaby, Gary. (2014). What's in a name? Place branding and toponymic commodification. *Environment and Planning A*, *46*(1), 153–167.
MEN. (2013, January 21). Boundary park bought back. Manchester Evening News. Retrieved from http://www.manchestereveningnews.co.uk/news/local-news/boundary-park-bought-back-1120835
Milligan, Melinda. (1998). Interactional past and potential: The social construction of place attachment. *Symbolic Interaction*, *21*(1), 1–33.
Millward, Peter. (2013). New football directors in the twenty-first century: Profit and revenue in the English Premier League's transnational age. *Leisure Studies*, *32*(4), 399–414.
Myers, Michael David. (2013). *Qualitative research in business and management* (2nd ed.). Los Angeles, London, New Delhi, Singapore, Washington DC: Sage.
Ogden, Mark. (2015, April 21). Manchester United rule out selling Old Trafford naming rights. The Telegraph. Retrieved from http://www.telegraph.co.uk/sport/football/teams/manchester-united/11594591/Manchester-United-rule-out-selling-Old-Trafford-naming-rights.html
Peck, Jamie, & Tickell, Adam. (2002). Neoliberalizing space. *Antipode*, *34*(3), 380–404.
Pike, Andy. (2011). Placing brands and branding: A socio-spatial biography of Newcastle Brown Ale. *Transactions of the Institute of British Geographers, NS 36, 206-222.*
Rose-Redwood, Reuben. (2008a). From number to name: Symbolic capital, places of memory and the politics of street renaming in New York City. *Social & Cultural Geography*, *9*(4), 431–452.
Rose-Redwood, Reuben. (2008b). "Sixth Avenue is now a memory": Regimes of spatial inscription and the performative limits of the official city-text. *Political Geography*, *27*(8), 875–894.
Rose-Redwood, Reuben, Alderman, Derek, & Azaryahu, Maoz. (2010). Geographies of toponymic inscription: New directions in critical place-name studies. *Progress in Human Geography*, *34*(4), 453–470.
Shoval, Noam. (2013). Street-naming, tourism development and cultural conflict: The case of the Old City of Acre/Akko/Akka. *Transactions of the Institute of British Geographers*, *38*(4), 612–626.
Stone, Chris. (2007). The role of football in everyday life. *Soccer & Society*, *8*(2–3), 169–184.
Taylor, Daniel. (2011, July 08). Manchester City bank record £400m sponsorship deal with Etihad Airways. The Guardian. Retrieved from https://www.theguardian.com/football/2011/jul/08/manchester-city-deal-etihad-airways
Tuan, Yi-Fu. (1974). *Topophilia: A study of environmental perception, attitudes and values*. Englewood Cliffs, NJ: Prentice Hall.
Vuolteenaho, Jani, & Ainiala, Terhi. (2009). Planning and revamping urban toponymy: Ideological alterations in the linguistic landscaping of Vuosaari suburb, eastern Helsinki. In Lawrence Berg & Jani Vuolteenaho (Eds.), *Critical toponymies: The contested politics of place naming* (pp. 227–251). Farnham: Ashgate.
Vuolteenaho, Jani, & Berg, Lawrence. (2009). Towards critical toponymies. In Lawrence Berg & Jani Vuolteenaho (Eds.), *Critical toponymies: The contested politics of place naming* (pp. 1–18). Farnham: Ashgate.
Vuolteenaho, Jani, & Kolamo, Sami. (2012). *Textually produced landscape spectacles? A Debordian reading of Finnish namescapes and English soccerscapes* (pp. 132–158). Helsinki: Helsinki Collegium for Advanced Studies.
Williams, John. (2006). 'Protect me from what I want': Football fandom, celebrity cultures and 'new' football in England. *Soccer & Society*, *7*(1), 96–114.
Woisetschläger, David, Haselhoff, Vanessa, & Backhaus, Christof. (2014). Fans' resistance to naming right sponsorships: Why stadium names remain the same for fans. *European Journal of Marketing*, *48*(7/8), 1487–1510.
Yeoh, Brenda. (2009). Street-naming and nation-building: Toponymic inscriptions of nationhood in Singapore. In Lawrence Berg & Jani Vuolteenaho (Eds.), *Critical toponymies: The contested politics of place naming* (pp. 71–84). Farnham: Ashgate.

Who owns the name? Fandom, social inequalities and the contested renaming of a football club in Timişoara, Romania

Remus Crețan

ABSTRACT

Recent scholarship in critical toponymy has raised questions about the impacts on urban residents of selling naming rights and the possibility of contesting the commodification of urban place names. This article examines these issues using Poli Timişoara, the major football team of Timişoara, Romania, as a case study. The cash-strapped local authority sold the naming rights for the team to raise revenue. However, when the club's private owner was unable to finance the team, the local authority reacquired the naming rights of Poli, now under the name Asociaţia Club Sportiv Poli (a new team with a small fan base). Drawing on the everyday narratives of fans, this article explores how the commodification of a local club name can cause social tensions. This is important for understanding urban residents' access, or lack thereof, to social capital resources and how this leads to the manifestation of social inequalities which co-determine fans' reactions and responses to the change in the name of a sports team. The findings suggest that commodifying names can mobilize serious contestations among partisan groups about who owns the name, although such contestations appear to be on uncertain ground because of post-communist urban practices of delayed economic restructuring.

1. Introduction

An increasing number of cities around the world are selling the naming rights of well-known landmarks, buildings, and other phenomena to businesses to generate additional revenue. Examples include the trading of naming rights of the First Gulf Bank Metro Station in Dubai, UAE and the 1–800-ASK-GARY Amphitheatre in Tampa, Florida (Rose-Redwood, 2011). The valuation process associated with naming rights is complicated by a variety of factors, such as new signage promoted under the umbrella of private-sector interests which are influencing and changing the naming process of the urban landscape (Light & Young, 2015). Consequently, an accurate evaluation of naming rights is required to reduce the conflict between public supporters of property and socially empowered figures in the business world. However, the (re)naming of sport clubs, including football clubs, is a relatively under-explored issue in critical toponymy. Edensor and Millington (2008, p. 190) argue that football clubs "are not placeless organizations. With their extensive networks of local and regional institutions, clubs have local roots and make material decisions that inextricably link or bind the football economy to particular localities." All sports clubs have working offices or headquarters and each has a designated geographical orientation (street name and street number). Consequently, the geographical location and the name associated with it provide a cultural background related to the activities of a particular place (Berg & Vuolteenaho, 2009).

The aims of this article are twofold: (1) to consider the responses of football fans to the sale of the name *Poli* (a football club in Timişoara, Romania); and (2) to consider how competing cultural claims are presented and managed when valuing the importance of local fans having access to social capital resources in an urban post-communist context. Exploration and analysis pursuant to these aims reveals how post-communist neoliberal practices are reframing the social scenes of contemporary urban groups through privatization of the spatial identities of football clubs, producing a range of social tensions as evidenced by the everyday experiences and narratives of fans.

This analysis makes an original contribution by examining responses to the sale of the name of a sports club, and the consequent inequalities among users of the major stadium in Timişoara; there are currently no extant studies of these issues in a post-communist context. As such, this research advances current debates about the sale of naming rights in critical toponymy (Light & Young, 2015; Rose-Redwood, 2011; Rose-Redwood & Alderman, 2011; Rose-Redwood, Alderman, & Azaryahu, 2010).

It is worth noting several important features to fully appreciate the context of the study area. Timişoara is a major city in the Western part of Romania, marking a point where the majority of foreign and domestic investment capital converges in the Banat region. It has a population of about 320,000, where cultural diversity is shared by 16 ethnic groups (Creţan & Matthews, 2016; Voiculescu & Jucu, 2016). In the history of Romania's football, several clubs from Timişoara, such as *Ripensia Timişoara* and *Chinezul Timişoara*, won the national championships during the interwar period (1919–1939). After the Polytechnic University and the team *Asociaţia Fotbal Club (AFC) Politehnica Timişoara* were created in 1921, and especially during the communist period (1947–1989) when Ripensia and Chinezul clubs were dissolved, the AFC team gained momentum and began to receive more attention from local fans. Furthermore, the construction of the *Dan Păltinişanu* stadium in the 1960s allowed the AFC team to move from the relatively small stadium of the Polytechnic University to a new 30,000 seat stadium (Figure 1). Following a complicated post-communist quasi-neoliberal path, AFC Politehnica Timişoara was relegated to the 5th division; the local authorities subsequently created a new team, *Asociaţia Club Sportiv (ACS) Poli Timişoara*, while most of the fans in the city embraced the idea of another new team – *Asociaţia Sportivă Universitară (ASU) Politehnica Timişoara*. The sequence of events have resulted in ongoing tensions among local fans, which are explored in further detail throughout this paper.

The remainder of the article is structured as follows. The second section examines the literature on selling naming rights and discusses debates connected to fandom. The third section considers the context behind changing of the *Poli* name, outlining the national and local nature of public–private practices and how these have impacted *Poli*

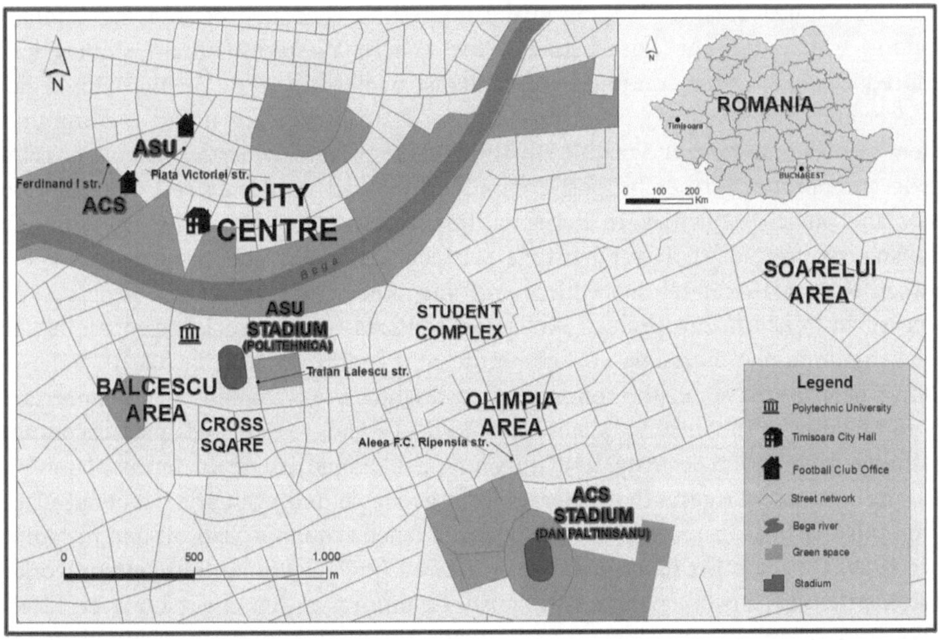

Figure 1. Study area: major football clubs and stadia in Timişoara. *Map implemented in GIS by Fabian Timofte, based on content provided by the author*

Timişoara. The fourth section introduces the methodology for the study. The fifth section explores results, delineated in terms of (i) identities, blame, and mistrust and (ii) varieties of post-communism through delayed and compromised economic restructuring. Finally, the last section presents the discussion and conclusion.

2. Selling naming rights, fandom, and social inequalities

Recent work in critical toponymies has identified a need for more engagement with the commodification of place-naming rights (Rose-Redwood, 2011; Rose-Redwood & Alderman, 2011; Rose-Redwood et al., 2010) or so-called *toponymy as commodity* (Light & Young, 2015). As the private sector's role in the transformation of the urban landscape has grown in the context of neoliberal urban governance (Hackworth, 2007; Ward, 2008), the selling of naming rights is one way for cash-strapped local authorities to generate revenue. Light and Young (2015, p. 441) argue that "there is a need for in-depth case studies of the commodification of urban place-naming rights. Such case studies could look at the practices of how city authorities and governing elites arrive at the decision to sell or auction their placenaming rights." This is why both institutional and legal mechanisms, along with the reasons behind the choice of private sector sponsors, deserve more attention. For all these challenges, it is reasonable to suppose that urban citizens (e.g., urban fans) would respond differently to the corporate renaming of public spaces in different political contexts.

In a Western European context, it is normal for football club sales to occur between private investors as opposed to local authorities; for those unfamiliar with the specificities of the Central and Eastern European context, things are different. Post-communist transformation has raised the interesting challenge of studying capitalist relations which did not exist before 1989. Where Western fandom studies have explored changes under conditions of capitalist ownership, clubs in the CEE region (and certainly in Romania) have moved from state ownership under a communist system into a complex and dynamic situation where private ownership is now possible; this is complicated further by post-communist delayed economic restructuring processes and other specificities. In Romania, this hesitant or rather compromised neoliberalism has created particular tensions between the public and private sectors, and between Romania's capital city (Bucharest) and other cities, which are generally not observed in Western European or American contexts. Within social sciences, debates on post-communist nostalgia for communism (Todorova & Gille, 2010) refer to positive perceptions about the communist past, which are a function of the uncertainties and failures of (neo)liberal democracy and the capitalist system i.e., macroeconomic instability, lack of job security, and burgeoning unemployment. In terms of football fans, the evidence suggests that nostalgia for the stable and egalitarian football in the communist past has increased with the extent of failed economic reforms and instability since 1989. On the other hand, there are critics on how the communist system worked in Romania and how this influenced post-communist entrepreneurship (see Stoica, 2004). Post-communist nostalgia, criticism on communism and the accumulation of current fans' frustrations and confusions have led to a post-communist crisis of fandom identity. For people who are deprived of a historical club name or of socialization on/around a long-established stadium, their claims may represent forms of restitution. Moreover, analysis of post-communist football fandom is almost completely absent in the current literature. In a recent article on the fandom of Dynamo Dresden, Daniel Ziesche (2017, p. 883) (2017) reveals how collective identity is linked to media images and how labelling and stigmatization could become "regimes of truth" for the identity construction of fans. In such circumstances, rationales and understandings pertinent to Central and Eastern Europe are sensitive because the context is different from that of Western societies.

It is surprising how little attention has hitherto paid been to the urban policy trend *vis-a-vis* football club naming rights, particularly given the bankruptcies that have occurred among football clubs in recent decades. Selling a club to a national or transnational company/owner and the trading of naming rights could be seen by some fans as a government strategy to increase overall revenue without increasing taxes (Bartow, 2007). Consequently, the selling of naming rights has emerged as an important issue in professional sports and this brings into attention the global-glocal debate (Swyngedouw, 2004), where networks, territories, and rescaling are at stake (Edensor & Millington, 2008; Hagen, 2011). While globalization emphasizes international influences, glocalization focuses attention on the interaction between the global and the local and the ways that this shapes governance and decision-making (Simonis, 2014; Swyngedouw, 2004). Moreover, glocalization relates to, and resonates with, the construction of place identity, in which fans' identities are influenced by the characteristics and cultural traditions of a given place. Some scholars have examined the changing spaces of football fandom in a global-glocal context (e.g., Bale, 2000; Conner, 2014; Evans & Norcliffe, 2016) although they tend to approach this issue from the perspective of Western clubs. In a case study of football fandom in the UK,

Evans and Norcliffe (2016) analyze how football spaces are economically and socially reproduced, and how supporters of two major teams (Everton FC and Liverpool FC) in the city work to both glocalize and globalize the football culture of Liverpool. Authenticity and "local superiority" are key issues of glocalization for the Evertonians, while the football spaces of Liverpool are more refashioned by global forces. Similarly, in a discussion of the history of Celtic FC, Conner (2014, p. 542) claimed that "the routes" – i.e., where the identities are being performed and (re) negotiated – are more significant in today's globalized world than the historical and cultural roots of these identities. The Celtic FC fan identification and attachment processes "are composed of a deterritorialized transnational group of individuals from around the globe who each have their own individual reasons for supporting the Club."

Caught between the global and the local, sports fans can strongly contest changes in naming rights of stadia and clubs. In many cases, the commodification of a name could lead to perceived social injustice and a lack of access to social capital resources. As a rather more limited concept in which we envision what moral society looks like, Harvey (1973, p. 97) defined social justice as "a particular application of just principles to conflicts which arise out of the necessity for social cooperation in seeking individual advancement." Thus, social justice refers to social and institutional arrangements related to the activity of production and distribution on an equal basis for all sections of the population and is a highly debated term in critical toponymies (see Alderman & Inwood, 2013). Conflicting claims could be resolved through commonly accepted principles of justice, although Harvey recognizes that "unfortunately, there is no one generally accepted principle of social justice to which we can appeal" (1973, p. 98). Therefore, *social inequalities* related to fans' access to social capital resources comprise unequal opportunities among the fans for a common benefit. In our case, we define social capital resources as "networks, norms, and social trust that facilitate coordination and cooperation for mutual benefit" (Putnam, 1995, p. 67). What Amartya Sen (2009) called an *inequality of capabilities* is, in fact, a result of what people can (or cannot) do with local resources (e.g., the power to participate in the social life of the community, access to sports, etc.) and this type of inequality seems to be critical for understanding fandom tensions created around the issue of naming (club) rights. Moreover, Antonio Gramsci's theory of cultural hegemony is relevant in arguing how dominant groups in a (capitalist) society create specific norms to suit their own purposes. Gramsci's theory acknowledges the economic and social constraints on the less powerful and allows us to analyze the systemic features of a society characterized by inequalities of power, where "subordinate groups lack the language necessary even to conceive concerted resistance . . . the capability for resistance flourishes and may lead to the creation of counterhegemonic alternatives" (Lears, 1985, pp. 573–574). Gramsci's cultural hegemony is not specifically about *manipulation*, but rather about *legitimation*. The experiences and values of dominant groups are validated in public discourse, while the ideas of subordinate groups are not, though it is a blurred line between dominant and subordinate cultures (Gramsci, 1971/1989). Gramsci's view of society involves a complex interaction of autonomous spheres of power (cultural, economic, and political; public and private) within a sum of practices or attitudes and not a mechanical model of superstructure. Thus, such a power "helps define the boundaries of common-sense "reality" either by ignoring views outside those boundaries or by labeling deviant opinions "tasteless" or "irresponsible" " (Lears, 1985, pp. 571–572).

While there is a dearth of academic literature concerning the changing of sports club names and stadia (see Leeds et al, 2007; Woisetschläger et al, 2014), the literature about the reactions of fans is equally limited. As Boyd (2000) appositely argued, arenas or stadia are physical memory places; but when commemoration is substituted for commercialization, then commodification becomes a threat to public memory places. The identity of memory places is thus altered because commemorative names of stadia are sacrificed for corporate naming. Taking Arsenal Football Club as an example, Church and Penny (2013, p. 830) highlighted the role of power and space seen in the new modern stadium as an *Arsenalization* of space. They concluded that through jointly agreed actions and contestation "supporters and stadium institutions not only negotiate but they will seek to influence practices and spaces." Others such as Woisetschläger, Haselhoff, and Backhaus (2014), contend that old stadium names could remain as sites of fans' resistance to naming right sponsorships. On the other hand, geographers have been interested in the relationship between branding football and local embeddedness, where "historical, geographical, cultural and economic embeddedness of football provides the ontological significance of the local" (Edensor & Millington, 2008, p. 188). In particular, match-day routines among fans of Manchester City FC have been transformed with fans preferring to socialize in central areas of Manchester and less around the stadium itself (Edensor & Millington, 2010). Pushing further the idea that the sociality of match-day rituals are evident nowadays for football fans in Norway and Britain, Hognestad (2012) argues that fans still relate to clubs as a consequence of physical considerations and social networking, even if television is considered to have the largest impact on the everyday community relationships of football fans.

In a comparable context to football stadia, football clubs are progressively purchased by wealthy individuals (and more rarely by multinational organizations) that exploit global football clubs (i.e., players, coaches, and administrators) for their benefit. This brings forth challenges experienced by the new owners when they deliver and consume social assets that are considered the property of the football business but also tokens of local identity. The transportability, as well as the lack of trust in new owners, is one of the main factors determining how fans evaluate the benefits and disadvantages of renaming a football club, and those fans may be willing to move their allegiances elsewhere. There are several meaningful comparisons of fans that moved entirely from one club to another location and (re) named it. The first is the formation of a new club by Manchester United fans who became disaffected with the corporatization of Manchester United and set up FC United of Manchester (Brown, 2008; Porter, 2015). In addition, there is the case of Wimbledon FC which moved to Milton Keynes and was renamed Milton Keynes Dons FC in 2002–2003; disaffected fans established a new club called AFC Wimbledon (Hognestad, 2012). Thus, professional sports clubs can move wholesale to different cities for financial gain, suggesting that "while some (football) brands may be global fluids, the economic, social and cultural embeddedness of most clubs in local contexts restrict the potential to extend brand awareness" (Edensor & Millington, 2008, p. 190). On the other hand, the transnationalization of the football economy (involving football clubs, the media, sportswear, the sports betting industry, security, the logistics and transport industry, and catering concessions) requires a portable workforce, consistently extending football marketing and branding (Medway & Warnaby,

2014). This scenario is particularly common but not exclusive to other institutions around the world, beyond those related to football clubs. For instance, the Roads and Transport Authority (RTA) of the Government of Dubai reported that a project to sell the naming rights of Dubai metro stations was "the 'ultimate branding and marketing opportunity' for national and international corporations seeking greater visibility in the fast-growing city" (Rose-Redwood, 2011, p. 35).

3. Selling of the poli name in a post-communist (football management) context

The change from communist political structures to a capitalist order has been an important period for people living in Central and Eastern Europe. For some people, post-communist nostalgia for the communist past was at stake (Todorova & Gille, 2010), while others were critics of communist state-led actions. Chelcea & Druţă (2016, p. 525) suggest that Central and Eastern Europe operates nowadays under the auspices of "zombie socialism" where "post-socialist transition is long over" and where political and economic elites in the region continue to use the ghost of state socialism as a disciplinary device and an "ideological antioxidant."

The landscape of political and economic transformation in Romania was based on uncertainty, high levels of unemployment, the privatization or closure of communist factories and businesses, and a hunger to attract new foreign direct investment. Romanian sports in general, and football in particular, cannot be divorced from these wider processes of transformation. In particular, the cash-strapped local authorities were intent on finding solutions for the major football clubs of the cities/towns to maintain football at high levels.

As the post-communist context of football management is inextricable from the case-study explored herein, it is pertinent to briefly introduce Romanian football and the ways in which its organization, governance, and management structures evolved in the post-communist period. *Federaţia Română de Fotbal* (The Romanian Football Federation; FRF) is the governing body for football in Romania. Its major role is to organize the national football team and all Romanian football structures and competitions. FRF has been affiliated with the *Fédération Internationale de Football Association* (FIFA) since 1923 and with the Union of European Football Associations (UEFA) since 1955. The major rules are administered by FRF's General Assembly and cover the status and transfer of players, disciplinary actions against football players and club managers, and the status of referees (FRF, 2017). Some of the duties of FRF were delegated to *Liga Profesionistă de Fotbal* (The Professional Football League; LPF), which is a sporting structure applicable to all professional clubs in Romania that are in the first football league – Liga I. In the period 1996–2013, the president of LPF was Dumitru Dragomir, a former communist manager of the team *FC Olt Scorniceşti*, Nicolae Ceauşescu's preferred team; the president of FRF was Mircea Sandu, a Romanian national football player in the 1980s. Both figures have been strongly contested by most football clubs in Romania, mainly by those outside the capital, Bucharest. Fans across Romania have accused them of being communists who are laundering money for the post-communist building *Casa Fotbalului* (The House of Football) and for their favoring of the principal clubs in Bucharest. Money laundering is a major reason why, in the post-communist period of Romanian football, some renowned clubs (e.g., *Rapid Bucureşti, Politehnica Timişoara, Universitatea Craiova*) were relegated to the 5th division. With the widespread prevalence of uncertainties (e.g., due to a lack of investors; corrupt so-called "investors" who launder the local authorities' money; clubs with debts to the state, etc.), many clubs have been renamed or have lost their naming rights, while others have even been dissolved.

Turning now to Timişoara, the club *AFC (Asociaţia Fotbal Club) Politehnica Timişoara* (widely known as *Poli Timişoara* or simply *Poli*) was established in 1921. It took its name from the city's Polytechnic University which initially ran the club. The club enjoyed considerable success – winning two Romanian cups and twice finishing as runner-up – and was supported by a loyal and passionate fan base. In 1999, an Italian investor (Claudio Zambon) acquired the club's naming rights and the team was subsequently renamed *FCU (Fotbal Club Universitatea) Timişoara* in 2004 (ACS Poli, 2017).

In 2008, a Romanian investor (Marian Iancu) was selected by city hall officials to invest in the club. However, Zambon still owned the naming rights for *FCU Politehnica Timişoara* so the team could not use *Poli* or *Politehnica* in its name. Hence, it was renamed *FC Timişoara*. In 2008, Iancu and the city hall officials regained the right to the name *Politehnica Timişoara*. Iancu was subsequently discovered to have been using the club for money laundering and was convicted (De Banat, 2014). As a result, in 2011, the Romanian Football Federation relegated the team to the 5th division; however, due to substantial debts, the club was eventually dissolved altogether. This relegation was a blow to the city's pride and consequently the newly-elected mayor (Nicolae Robu) took action to restore first division football to the city. City hall officials had two options: re-start from the 5th division (the lowest division in Romania's domestic league) or merge with a football club from the higher divisions for quick results. Robu took the latter course and made an arrangement in 2011 with the nearby town of Recas to rename *ACS (Asociaţia Club Sportiv) Recaş*, a local team which had been newly promoted into the 2nd division, into *ACS (Asociaţia Club Sportiv) Poli Timişoara*. All the players from ACS Recaş transferred to the new team, leaving the town of Recaş without its own football team. The rights to the name of the historical *Poli* (along with its colors and flags) were acquired by the Timişoara town hall for *ACS Poli Timişoara*. This team was promoted to the 1st division in 2012 (ACS Poli, 2017).

While *ACS Poli Timişoara* is the team with the official support of the city hall and county authorities, it enjoys little support among fans. The creation of *ACS Poli Timişoara* outraged thousands of historical *Poli* fans who deeply distrusted the actions of the city hall. Fans regarded *AFC Politehnica Timişoara*, which was relegated to the 5th division in 2012, as the "real" *Poli* team; they resented the fact that the original *AFC Politehnica Timişoara* was not given the opportunity to work its way up from the 5th division on its own merits. Distrustful of private investors, a group of fans and sponsors established a new non-governmental organization in 2013 and adopted the team name *ASU (Asociaţia Sportivă Universitară) Politehnica Timişoara* (ASU Politehnica, 2017). The team is now in the 2nd division. It plays in a small stadium in the city but regularly attracts 30,000 fans. Meanwhile, *ACS Poli Timişoara* plays in one of the largest stadia in the country but with crowds of 800 people or less (Figure 1).

As noted above, *ACS Poli Timişoara* is the team that has the support of the county seat and the local town hall authorities. In an interview with a former local councilor associated with *Poli Timişoara* (who requested to remain anonymous), we found that "the name was obtained from the National Register of Federations, Foundations, Association and Trade. To avoid confusion…the name *Poli* in *ACS Poli* was selected rather than Politehnica." The formation of the team resulted in the addition of ACS due to the connection with ACS Recaş; this allowed a fusion of the two teams, whilst also preserving the initials. Moreover, the local councilor stated that "anyone who wants to show an association with *Poli Timişoara* will have to make mention of the name *ACS Poli Timişoara.*"

A number of organizations have used the name Poli, but the press office of the Romanian National Registry of Associations, Foundations, and Federations explained that the value and meaning of these names differ on the basis of what initials or words are used with it. Despite the doubt that surrounds the legitimacy of *ACS Poli Timişoara* being the original team, the local state has heavily funded this team, meeting most of its expenses with sponsorship deals. Meanwhile, fans of the "other" Poli team, *ASU Politehnica Timişoara,* have resisted by using the slogan "Poli we are" for the team, which also wears white and violet uniforms. Trams and buses in Timişoara are all colored in white and violet as a local symbol. A former staff member *of ACS Poli Timişoara* stated, "The current regulation of FRF does not allow us to financially support both teams [ACS and ASU]...similar or identical names are not the norm; we can take note of ASU's affiliation to FRF, but the historical sign, colors and flag [belong] to ACS."

Despite popular support for *ASU Politehnica Timişoara,* major corporate sponsors (e.g., Romanian domestic furniture producer *Casa Rusu,* automotive group "Continental, etc.)" and authorities continue to support ACS in every respect (Monitorul, 2016). The difference between the funding available and expenditure of the two teams is substantial. The political favor given to the ACS club to start from the 2nd division is a depiction of the social inconsistencies associated with the naming issue. Fans' opinions had been ignored by the authorities supporting ACS due to the economic advantages they received from sponsors; the authorities perceived the actions of the die-hard ASU fans as unrealistic. Hence, fans stepped up to support the emerging ASU team and collected funds by themselves to sponsor it for the sake of preserving traditions and values.

4. Methodology

The methodology adopted for this case study follows the pattern of other case studies in social sciences (i.e., Patton, 2000; Yin, 2009) and more specifically in human geographies (i.e., Dunn, 2005). The paper is framed exclusively as a case study which is not necessarily generalizing results. The benefits and strengths of case study research are significant. First, although the force of transferability and the example are often underestimated for case studies, they can be useful for generating and testing a hypothesis, and for disproving of preconceived notions rather than theory verification (Yin, 2009). Second, they offer insights into real life practices; one can have more learnings stemming from these intense observations than by a study involving only statistics of a large group (Holt-Jensen, 2013, p. 9). I use this case study to throw light on the commodification of names and the implications of this process in the context of post-communist CEE.

The connection between the three main elements of the paper (naming rights, fandom, and social inequalities) is emphasized as a key issue. Interviews with local officials and fans of both local teams (ACS and ASU) provide the primary data for this study, augmented by data from secondary sources (newspapers, national and local football websites).

The sampling strategy for interviewing the fans was as follows: 10 of the author's friends who are fans of both teams were contacted. These people were used as the basis for snowball sampling (relying on referrals from initial subjects) to generate further interviewees, rather than sourcing participants on social media. Snowballing has advantages over using social media to source participants because it is more efficient in securing interviewees who are rich in information or learning about critical cases (Patton, 2000). Indeed, the snowballing approach proved to be useful; while various potential interviewees were recommended initially, it became apparent that certain names were mentioned over and over again. In other words, there was inter-subjective agreement about key people who should be approached for interviews. Thus, it can be asserted that the interviewees for the study provided a reasonably salient sample.

Interviews were conducted in August 2016 with 120 fans from both sides collectively: *ACS Poli Timişoara* (n = 52) and *ASU Politehnica Timişoara* (n = 68). See Table 1 for broad characteristics of the sample of fans.

In designing and carrying out the interviews, care was taken to avoid cognitive burden: participants needed to understand the concepts of the research (e.g., neoliberalism, naming rights) but this was facilitated by circumventing specialist terminology, where necessary and possible, in favor of more everyday language. Furthermore, not everybody could discuss their understandings and opinions with ease which posed complexities in terms of ensuring equal elicitation opportunities among interviewees. As such, a balance between speaking and listening, watching body language, remembering (Cloke et al., 2004) as well as paying particular attention to limited degrees of reflexivity was important.

Interview topics included: resistance to the naming rights of *Poli*; connections with protestors in other cities/countries; the availability of website platforms (e.g., blogs, online journals, and comment rooms); the role of private owners in selling the name of *Poli*; and fans' everyday connections to social capital resources. Further, local officials

Table 1. Interviewee characteristics – Football club, age, gender, and profession.

	18 and under		19–64		65 and over		
	Female	Male	Female	Male	Female	Male	TOTAL
ASU Poli Fan	10	4	18	31	2	3	68
ACS Poli Fan	5	5	11	26	1	4	52
Profession							
High-School Student	15	9					24
Unemployed			4	8			12
Public Sector Worker (Non-Manual)			5	11			16
Public Sector Worker (Manual)			5	12			17
Private Sector Worker (Non-Manual)			6	8			14
Private Sector Worker (Manual)			2	6			8
Third Sector Worker (i.e., person who work for non-governmental organizations including charities and voluntary organizations)			7	12			19
Pensioner					3	7	10
TOTAL	**15**	**9**	**29**	**57**	**3**	**7**	**120**

were interviewed to identify, clarify, and explore relevant procedural and legislative issues. These included a former staff member of ACS Poli Timişoara and a former local city councilor who had been thus involved in the decision-making process of re-naming and selling the club's name.

All interviews were conducted in Romanian. To encourage participants to discuss and represent their perceptions, a narrative (rather than structured) note-taking (rather than audio recording) approach was adopted. I was aware that the participants' identities impacted their responses, mainly given participants' gender framework. As most of the interviewees were male (Table 1), who proved to be more dedicated to the faith of the local teams and more affected by the tensions in the selling of the *Poli* name, I consider that the results were appropriate and showed a general picture of the current fandom positions in Timişoara. Moreover, most of the interviewees were educated young people (high-school and university students; intellectuals, such as doctors and teachers, as well as people who work for non-governmental organizations), which reflected in the quality of the collected data.

The analysis of secondary sources was based on selecting major editorials dealing with debates on the *Poli* name. Following interpretation methods of primary and secondary sources in social sciences (Dunn, 2005; Patton, 2000), a table with values (themes and quotes) was created for depicting the necessary data. Then major themes were identified and relevant actors' quotes were chosen. The results were explained and connected to the examination of the sale of *Poli* name.

5. Results

5.1. *Identity, blame, and mistrust*

Fandom has long been an intense aspect of Timişoara culture. The resistance of ASU fans began after 2012 in conjunction with the name change, and their frustration is represented in an interview with an ASU fan club leader: "I understand they need investors, but they left us to continue our forefathers' will [heritage] from the last hundred years." Considering both the views of fans who contested the *Poli* naming rights and fans of ACS (who did not), all the interviewees were clearly influenced by the construct of place identity, which suggests that people's self-identity is influenced by the characteristics of a given place (Hauge, 2007). They were also well informed about the meaning of naming rights, but each side had their own opinions.

The main targets of the resistance were the local politicians who were responsible for the divided situation. All the fans interviewed questioned the motivations of politicians, including the current mayor and local town hall councilors, whose links with wealthy investors were raised as a matter of concern. Many participants expressed mistrust of their local leaders because they wasted monetary resources and offered little support to the community. Others echoed this point, noting that investors were as much to blame as politicians. A central issue that drove the protests of ASU supporters was the loss of income and access to the *Dan Păltinişanu* stadium resulting from the change in leadership. A lack of consideration for the fate of the team and players' livelihoods was also mentioned. One interviewee commented, "we had to move from our historical stadium" and "I paid a lot of money as a public contributor to the local budget, but have nothing to show from that." The fans perceived that local leaders were untrustworthy and had spent local resources on a product that was irrelevant to their needs.

The Romanian Football Federation and EU Football Courts were also blamed because regulations produced by FRF in Bucharest were imposed on clubs throughout the country. The fans expressed mixed feelings about the impact of the domestic federation on their practices. Some argued that FRF contained corrupt individuals who were not interested in developing regional and local football traditions, with an ASU interviewee claiming, "The Romanian Federation put pressure on us to respect their laws…they wanted to destroy regional brands like *Poli* because it is an old club and they were interested only in developing their own affairs, laundering public money…most of the former and current leaders have dozens of legal problems; more than that, some of them are in jail." The perception that FRF wanted to maintain control, regardless of cultural and historical traditions, was widespread. The rivalry with Bucharest has a long historical background. Tensions exist between "the periphery" of the current study (Timişoara city and the Banat region) and the capital city of Bucharest. Here it is important to briefly explain the cultural tensions between people living in the Banat region and Bucharest. Located at the edge of the Pannonian Plain, Banat includes part of western Romania, northern Serbia, and south-eastern Hungary, with Timişoara being its largest city. As a genuine multicultural region made of 16 ethnic minorities, people are proud of their Habsburg and Austro-Hungarian cultural influence (see Creţan, Turnock, & Woudstra, 2008, pp. 17–19). On the other hand, Bucharest, situated in southern Romania, was part of the historic province of Wallachia and was under the influence of the Ottoman Empire for more than five centuries. A long-standing distrust still lingers toward Bucharest, where corruption is considered by Timişoara fans as a norm among politicians and football leaders. These tensions could be conceptualized as a form of regional competition.

Interviewers noted that the mismanagement eliminates opportunities for sources of income, such as revenue from tickets, or leads to *divide et impera*, where fandom from different historical Romanian clubs "are now divided and relegated to the 4th or 5th divisions." Interviewers provided examples of other Romanian teams facing disparities in treatment. One commented, "see what happened…to *Steaua Bucureşti* and *Universitatea Craiova*," citing another case of perceived inequality. The teams cited by the interviewees, Steaua Bucharest and *Universitatea Craiova*, are also afflicted with divided fans. Steaua is the former army team, but the Ministry of Defense refused to sell it to a wealthy investor, George Becali. As a result the dispute over naming rights resulted in two teams: one with the right to the name *Fotbal Club Steaua Bucureşti* (in the 4th division) but with few fans, the other entitled Fotbal Club FCSB (with an extensive fan base) which is currently in the 1st division. Similarly, most of *Universitatea Craiova's* fans joined the newly created local authority team, *FCU Universitatea Craiova*, while a few supporters fused with the 4th league relegated *Fotbal Club Universitatea Craiova*. The situation was further skewed in this direction after the local municipality finalized the construction of a 40,000 seat stadium for FCU in 2017. Accordingly, the Timişoara case is not unique in the context of post-communist Romanian fandom.

Although EU football bodies and FRF were blamed for regulating the club's naming rights, interviewees did not indicate awareness of similar resistance or dissatisfaction in other European countries. The costs and risks associated with resistance suggest other non-contentious channels were not useful in challenging the loss of naming rights and the rights of ASU team to play at the *Dan Paltinişanu* stadium. This is the largest stadium in Timişoara (and the largest in Romania at the time of construction in 1963, with 30,000 seats) placed near the Student Campus area of the city and bearing the name of a *Politehnica Timişoara* player from the 1970s. Asked whether they had attempted to use regular channels in their resistance, most participants mentioned they had used Facebook and ASU's blog to complain about the local authorities, but to little effect. One ASU fan stated, "We, the fans, are very united. If we hear of [the authorities] limiting our rules and rights, then we have to protest. But our voice is only reflected in the local media, while the authorities have done nothing for us. We wrote on many walls in the town that 'AC Recaş is not Poli' but the authorities did not care about us." Even the Polytechnic University's Rectorat building in the central area of the city was inscribed with this graffiti (Figure 2) because the current mayor was the former rector of the University (2004–2012). These comments suggest that the influence and usefulness of official representative bodies is perceived to be limited, with one participant grouping their leaders alongside members of the local council and the mayor as being responsible. Reliance on a shared identity as Banat people was a key aspect of fandom and provided a mechanism through which they could organize and mobilize.

The glocalization of ASU fans works both at the historical heart of the game and in local territories that have adopted football; for instance, if we compare the supporters of Liverpool (Everton FC and Liverpool FC) with those of Timişoara we can notice that the fans of ASU perceive the "localness" of the support base as a "virtue," exactly as in the case of Everton FC supporters in Liverpool (see Evans & Norcliffe, 2016, p. 226). As noted, culture and tradition were more powerful forces for the ASU fans, potentially enabling the establishment of dense networks of trust. However, one ACS fan said, "I

Figure 2. Form of contestation on the wall of Polytechnical Rectorat building in the centre of Timişoara.

usually counter-write on the city walls that 'AC Recaş nu e Poli' ['AC Recaş is not Poli'] with 'ACS is Poli' because, for me, the team who plays at *Dan Paltinişanu* stadium is the real *Poli*, and this is ACS *Poli*." Therefore, resistance came from both sides in a fight and counter-fight for claiming the rights to the name of *Poli*.

5.2. *Varieties of post-communism through delayed and compromised economic restructuring*

The loss of a historical name was clearly foreshadowed by events prior to the 1990s. Indeed, the role of communism and, more exactly, the move away from communism, was an important part of the context for the sale of (and response to) the name *Poli*. As for the implications of the end of communism and the particular circumstances of post-communism in Romania and how they affected the way that the naming issue of *Poli* unfolded, several political and economic stages can be traced: instability (early 1990s); reformist government (1996–2000); return to power of former communists (2000–2004); Romania joining the EU (2007); global financial crisis (2008–2011); and most recently recession and frequent public protest (2012 onwards).

First, the political and economic instability of the early 1990s was associated with former communist elites in power (Boia, 2001). This period of uncertainty led to the phenomenon of nationalism (see the anti-Hungarian events in Târgu Mureş in March 1990) and the *mineriade* (i.e., miners called by left-wing president Ion Iliescu to come to Bucharest to defend his position in front of protesters on June 1990 and September 1991). Moreover, statements by the president such as "*nu ne vindemţara*" ("we don't sell our country") reflected the former communist elite's distrust of neoliberalism, and a distaste for foreign investors. In fact, it was a gradualist approach to reform, characterized by economic decline and recession, but little actual restructuring. This atmosphere led to falling living standards, the rise of ultranationalism, and a deterioration of Romania's international image. This general political background strongly affected football support. Some traditional and local football teams had to reorganize on the basis of private Romanian investment because foreign investors were still seen as "enemies" of the country, while former big communist enterprises underwent restructuring (see Ilieş, Ilieş & Murariu, 2016). Moreover, some football supporters and teams became associated with ultranationalist parties.

Second, the government of the 1996–2000 period had lofty ambitions but achieved only limited actual reform. That government continued the process of limited privatizations, while living standards continued to fall (Boia, 2001). Interestingly, many communist industries experienced bankruptcy. In terms of the mid- and late-1990s market reforms, Stoica (2004, p. 274) highlighted that "in Romania most medium-size and large enterprises are still owned by the state, and the several attempts at mass privatization have been unsuccessful." In this context of lack of private investors and collapsing living standards, football support continued to decline. Nevertheless, mayors and local government officials still found resources to gain electoral capital through building new stadia and paying football players.

Third, when former communists returned to power (2000–2004) a slow but steady overall consolidation of the Romanian economy was noticeable, with economic growth and marginal increases in living standards as well as an increasing engagement with joining of NATO and the EU. Generally, this helped improve Romania's international image. In these conditions, football teams and their supporters benefitted from attracting foreign investors and football players.

Fourth, after Romania joined the EU in 2007, football teams enjoyed increased stability and increased foreign investment due to political and economic stability at the national level. The massive change in mobility (with Romanians freely travelling within the EU and people from other countries visiting Romania) enhanced the potential to attract team funding and football players from other EU countries. The increased living standards and the massive exposure to "European" ideas and models positively affected football teams and their supporters. As Young and Light (2001, p. 953) argued, "[CEE] states are often represented as 'Western' in a process of 'Europeanisation' which seeks to emphasise their modern, liberal, democratic nature." Foreign ownership became a norm for more football teams in Romania; however, because Romania was not immune to the deleterious impacts of the global financial crisis of 2008–2011, the neoliberalization of teams was not fully consolidated.

Fifth, the post-2011 economic recession and the resultant demonstrations (see for instance Creţan 2015; Vesalon & Creţan 2013) was another barrier to football development and management. The events led to falling living standards and had further impacts for both football players and teams.

Summing up, broader political and economic post-communist developments which may appear to be completely unrelated to football strongly affected what happened to football teams and their supporters. This "big picture" of limited and hesitant economic restructuring and post-communist political/economic shifts also led to changes in the everyday behaviors of football teams and supporters at the national level and manifested itself at the local level with Timişoara's football teams.

Among the fans interviewed, there were mixed feelings about the 1990s. Most felt that the period after communism failed to achieve anything and that they were financially worse off as a result. Some claimed that the FRF regulations were, in fact, worse than the regulations of the communist period. This is connected to post-communist nostalgia. One participant argued, for instance, "The new political system was difficult to be understood; clubs could not be financed from local budgets, and they had to attract foreign investors [...] but it seems most of them were not thoroughly vetted to confirm if they really had money-making potential, and usually the process bankrupted the clubs." It was also noted that foreigners and Romanians with small businesses in Timişoara were allowed to finance the team for a short period of time, but local authority leaders did not have a vision for long-term sustainable investment. In the past, the absence of private ownership had made the continuation of historical practices and names easier to maintain. At the same time, corruption associated with former members of the Communist Party, who were involved in leading the club, had also caused frustrations. Interestingly, the Romanian interviewees have very different attitudes about socialist and communist systems. Those unfamiliar with the systems – mainly the younger, post-communist generation, incorrectly consider them inseparable, while those in various academic disciplines, such as economics and sociology, tend to focus on aspects of the systems most pertinent to their own work. Thus, socialism system relies on the idea that capitalism can exist in a socialist state and capitalism can work in a socialist state; meanwhile, communism is based on the principle that capitalism must be destroyed in order to develop a classless society. If communism is both an economic and political system, socialism is primarily an economic system.

Almost none of the fans had heard of neoliberalism, but their comments can be interpreted through that lens. Both ASU and ACS fans agreed that football cannot survive without private owners and cash subsidies, but some of them went further: one ASU fan declared "we don't need corrupt local authorities to finance the club. We don't need tricksters [under the guise of private ownership] to come here and launder money. Now our club is in the 2nd division thanks to the fans' financial contributions - everything is transparent, no private thieves, no local leaders involved."

The issue of losing rights to the *Poli* name had significant impacts, as it resulted in the redirection of local resources in the city and, thus, a loss in social capital. All ASU participants expressed their frustration at the need to gather in a new small stadium. Moreover, networks of personal contacts were destroyed. Many fans had common chat rooms with the name *Poli* Timişoara, and now they had to create a new website named *ASU Politehnica*. They had to break long-standing friendships with fans that were then neutral or who transferred their allegiance to ACS. An ASU fan commented, "The relationship with my son has been tense since he became an ACS fan," and "I lost many friends who went with the mayor [i.e., the ACS team]." Losing the opportunity to network with other fans was also a major loss for one ACS fan who noted, "When we were a single club in the city, we had very good channels to meet each other...a simple Facebook announcement on the *Politehnica Timişoara* Facebook page and we would gather at McDonald's and march to the stadium. Now we are [a] few fans at ACS and not united...Nobody sings and shouts for our team, it is like at the theatre - silence. It's a pity we have no gallery, no strong networks to meet and build a well identified fandom." Many ASU fans are still active in trying to establish a vibrant atmosphere at the stadium. One fan stated, "Even when we were in the 5th division, we went with buses and personal cars to different villages in Timiş County where the team played, chanting, waving our white and violet flag."

Considering the diverse views expressed in the interviews, it is clear that *Poli's* traditions feature strongly in their concerns. Although the participants argued that threats of losing ownership of the *Poli* name mobilized the anger of fans, there was a shared identity and perception in terms of recent events. Poor management by FRF and local authorities were perceived as undermining established ways of life, even if they did not lead to frequent mobilizations. The issue of inequality was mentioned by almost all ASU participants, noting that privatization of the club and the Romanian Federation's attempts to control behavior presented a significant challenge to the continued existence of the team. Reference to forefathers and how they managed the team appeared repeatedly in the interviews, reinforcing the strength of tradition, again suggesting post-communist nostalgia for the more secure and safe system of Romanian football under state communism. There was an overriding sense that they wanted to continue with their historical name. Many participants noted that limited access to social capital resources was a key issue in producing social inequalities mainly due to the local administration decision on imposing a lack of access to *Dan Păltinişanu* stadium for the ASU supporters. This further strengthens the perception that fans are increasingly losing rights, including naming rights, under neoliberal pressures.

6. Discussion and conclusion

This study sought to evaluate the impacts of fandom and its social implications in the type of spatial conflicts emerging between the ownership of a sports club and the local communities involved, pushing forward the current debates in critical toponymies on naming rights. There have been recent calls for more study of the commodification of place-naming rights (see Light & Young, 2015; Rose-Redwood, 2011; Rose-Redwood & Alderman, 2011; Rose-Redwood et al., 2010); this article demonstrates that changing the name of a sport team is an arena of contestation. The legitimization of the ACS club name by the local city authorities as *Poli* was strongly opposed by fans, who believed that the name violated the historical values of the team and club that the fans associated with the spirit of the game. Although acceptance of naming deals can alleviate clubs' financial problems, this case-study illuminates how a backlash from fans can have substantial negative impacts on the perception of that club within local communities. As Woisetschläger et al. (2014) argued, by viewing the resistance of fans through the lens of social identity theory, sponsorship associated with naming rights can be conceived as an act of out-group influence on the ritual place of the in-group. Hence, while evaluating the impacts of sponsorship due to name changing, researchers need to account for the responses of fans. In this case, that response was powerful, as the fan base rejected the decision of the local authorities and established their own team which was prepared to work its way up from the bottom.

The inequalities existing in the funding offered to the two teams claiming to be descendents of *Poli* (or *Politehnica Timişoara*) are evident. The preference of those in power for ACS left the ASU team – which enjoyed the support of the majority of fans – without any official funding or support. Antonio Gramsci's cultural hegemony helps us to better understand how dominant groups in society "create" norms, especially in relation to economic contexts, to suit their own purposes. As Lears (1985, p. 573) puts it: "People indeed create their own symbolic universes (Gramsci's spontaneous philosophy) to make life understandable and tolerable, and those symbolic universes do come to have an apparently 'objective' validity, particularly over generations as they spread from scattered individuals to broad social groups." A given symbolic hegemonic universe can serve the interests of dominant groups to the detriment of others (Gramsci, 1971/1989). For instance, ACS fans are typical of subordinate groups who legitimize dominant voices, sharing a kind of complicity in their own victimization. On the other hand, the fans of both teams were convinced that FRF was irresponsible in the way that it implemented regulations that appeared to favor the interests of teams in Bucharest. Some of the comments of fans also suggested that post-communist nostalgia may have affected their attitudes, although those unfamiliar with socialism and communism systems incorrectly consider them synonymous.

Football is a key area of pride in local communities, recognized by fans through their ways of celebrating local identities, traditions, and myths (Edensor & Millington, 2008). Although the sports industry has transformed itself in the past few decades, shared identity and glocalization remained very strong. Former major clubs, such as

Politehnica Timişoara, are symbols of fans' regional/local social and cultural identities rooted in a tradition of rule by the Austro-Hungarian Empire, as opposed to those in Bucharest which are considered by the local fans to be more Balkan in style. In this sense, the current attitudes of ASU fans reflect a wider division within contemporary Romania between Bucharest (with its history of Ottoman rule) and Transylvania/Banat (with its history of Austro-Hungarian rule).

The case study presented in this paper is novel relative to the extant literature, for a number of reasons. First, it explores the response of fans to a change of name, highlighting elements of social inequalities and injustice that exist in the preferences shown by officials in power; in this case, the inequality was reflected in the fact that fans who opposed ACS and supported ASU had no access to capital and had to raise money on their own to support their club. Second, this study is rooted in a specific context – post-socialism – which has been characterized by intricate political and socioeconomic issues of transformation that appeared due to the shift from the communist era to a capitalist system (Stoica, 2004; Young & Light, 2001). As the former communist cadres (who did not have experience as entrepreneurs) were at an advantage in being employers due to their organizational experience and network resources established since the communism period (Stoica, 2004), post-communist Romania was characterized by a quasi-neoliberal reform, which led to compromised economic restructuring. On the other hand, local political elites were eager to rapidly identify and capture venue sources to support football clubs and put their trust, often misguidedly in investors in order to gain popular support and achieve electoral success. However, this resulted in many major football clubs in Romania becoming bankrupt and even being dissolved. In addition, corruption and money laundering undertaken by leaders of FRF and LPF, as well as by local mayors and private investors, gave birth to sorrowful examples like that of *Poli Timişoara* fans. This kind of "zombie socialism" (Chelcea & Druţă, 2016), defining the post-communist period, is an effect of national political and economic restructuring which delayed/compromised economic restructuring generated in the post-1989 period.

Doreen Massey (1991, p. 29) explained that articulated moments in networks of social relations are necessary for analyzing spatial relations. Several such moments were revealed in the interviews for this case study in connection with access to social capital. Local communities have a right to networking and to the capital budget that is allocated for local sports teams. However, when local communities have limited access to capital resources, then the changing of naming rights can give rise to the perception of social injustice. Although conflict and rivalry in football are noteworthy everywhere, bonds of mutuality and (transnational) networks continue to be the most important issues that enable football fans be in contact with one another (see also Hognestad, 2012).

In summary, while drawing on the everyday experiences and narratives of fans, this article shows how the geographies of commodifying the name of a local football club are relevant in the manifestation of social tensions and inequalities. This is important for understanding local people's access to social capital, including networking, resources, and support from local leaders. The findings push forward current theories of critical toponymies by highlighting that commodifying names can mobilize serious contestations between partisan groups about who owns a name; such contestations, however, appear to be on uncertain ground because of post-communist delayed economic restructuring practices.

Further research could explore the relationship between naming rights and football clubs' social and economic behavior to better understand if – and how – fans are respected and social justice is served. There is also ample scope to explore the aspects of fandom that are significant to place-based cultural and historical attachments with clubs. The owners of football clubs together with local decision-making actors need to be aware of such attachments when considering the sponsorship of a team's name (or even a change of name) since such proposals can trigger massive levels of resistance and contestation from the fans themselves.

Acknowledgements

I am grateful to the editors of the special issue on "Naming Rights, Place Branding and the Cultural Landscapes of Neoliberal Urbanism," especially to Duncan Light and Reuben Rose-Redwood who provided important suggestions on the article. I would also like to thank Susan Moore and the three anonymous reviewers for their helpful comments on earlier drafts. The map included in this paper was thoroughly edited in ArcGIS by Fabian Timofte.

Disclosure statement

No potential conflict of interest was reported by the author.

ORCID

Remus Creţan http://orcid.org/0000-0002-9053-8394

References

ACS Poli. (2017, March 6). *Despre noi: Echipa şi conducere (About us: The team and management)* Retrieved 25 July 2016 from http://www.acspoli.ro/index.php/conducere/

Alderman, Derek H., & Inwood, Joshua. (2013). Street naming and the politics of belonging: Spatial injustices in the toponymic commemoration of Martin Luther King Jr. *Social and Cultural Geography*, *14*(2), 211–233.

Bale, John. (2000). The changing face of football: Stadiums and communities. *Soccer & Society*, *1* (1), 91–101.

Bartow, Ann. (2007). Trademarks of privilege: Naming rights and the physical public domain. *UC David Law Review*, *40*(1), 919–970.

Berg, Lawrence D., & Vuolteenaho, Jani (Eds.). (2009). *Critical toponymies: The contested politics of place naming*. Farnham: Ashgate.

Boia, Lucian. (2001). *History and myth in Romanian consciousness*. Budapest: Central European University Press.

Boyd, Josh. (2000). Selling home: Corporate stadium names and the destruction of commemoration. *Journal of Applied Communication Research*, *28*(4), 330–346.

Brown, Adam. (2008). 'Our club, our rules': Fan communities at FC United of Manchester. *Soccer & Society*, *9*(3), 346–358.

Chelcea, Liviu, & Druţă, Oana. (2016). Zombie socialism and the rise of neoliberalism in post-socialist Central and Eastern Europe. *Eurasian Geography and Economics*, *57*(4–5), 521–544.

Church, Andrew, & Penny, Simon. (2013). Power, space and the new stadium: The example of Arsenal football club. *Sport in Society*, *16*(6), 819–834.

Cloke, Paul, Cook, Ian, Crang, Philip, Goodwin, Mark, Painter, Joe, & Philo, Chris. (2004). *Practicing human geography*. Thousand Oaks, CA: Sage Publications.

Conner, Neil. (2014). Global cultural flows and the routes of identity: The imagined worldsof Celtic FC. *Social & Cultural Geography*, *15*(5), 525–546.

Creţan, Remus, (2015). Mapping protests against dog culling in post-communist Romania, *Area*, *47*(2), 155–165.

Creţan, Remus, & Matthews, Phillip W. (2016). Popular responses to city-text changes: Street naming and the politics of practicality in a post-socialist martyr city. *Area*, *48*(1), 92–102.

Creţan, Remus, Turnock, David, & Woudstra, Jaco. (2008). Identity and multiculturalism in Romanian Banat. *Méditerranée*, *110*(1), 17–26.

De Banat. (2014, October 14). *Marian Iancu, condamnat la 12 ani de inchisoare cu executare*. Editorial by Vlad Stoian. Retrieved 25 July 2016 from www.debanat.ro

Dunn, Kevin. (2005). Interviewing. In Iain Hay (Ed.), *Qualitative research methods in human Geography* (2nd ed., pp. 79–105). Oxford: Oxford University Press.

Edensor, Tim, & Millington, Steve. (2008). 'This is our city': Branding football and local embeddedness. *Global Networks*, *8*(2), 172–193.

Edensor, Tim, & Millington, Steve. (2010). Going to the match: The transformation of the match-day routine at Manchester City FC. In Sybille Frank & Silke Steets (Eds.), *Stadium worlds: Football, space and the built environment* (pp. 146–162). Abingdon: Routledge.

Evans, Daniel, & Norcliffe, Glen. (2016). Local identities in a global game: The social production of football space in Liverpool. *Journal of Sport & Tourism*, *20*(3–4), 217–232.

FRF (Federatia Romana de Fotbal). (2017, February 6). *Despre FRF: Statut si Regulamente (About FRF: Status and Laws)*. Retrieved 25 July 2016 from http://frf.ro/statut-si-regulamente.html

Gramsci, Antonio. (1971/1989). *Selections from the prison notebooks*. (Quentin Hoare & Geoffrey Nowell Smith, Eds. & Trans.). New York: Reprint.

Hackworth, Jason. (2007). *The neoliberal city: Governance, ideology, and development in American urbanism*. Ithaca, NY: Cornell University Press.

Hagen, Joshua. (2011). Theorizing scale in critical place-name studies. *ACME: an E-Journal for Critical Geographies*, *10*(1), 23–27.

Harvey, David. (1973). *Social justice and the city*. Oxford: Basil Blackwell.

Hauge, Åshild L. (2007,March 22). Identity and Place: A Critical Comparison of Three Identity Theories. *Architectural Science Review*, *50*(1), 44–51. Retrieved 23 July 2016 from http://faculty.arch.utah.edu/benham/group%203/Place-Identity.pdf

Hognestad, Hans K. (2012). Split loyalties: Football is a community business. *Soccer & Society*, *13* (3), 377–391.

Holt-Jensen, Arild. (2013). *The case-study approach: How to design a good interview and make arrangements for interviews*. Oslo: Nord Plus.

Ilieş, Alexandru, Ilieş, Marin, & Murariu, Călin. (2016). Socialist heritage and symbols in football teams (1981-1989) in maramureş county (romania). *Geojournal Of Tourism and Geosites*, *18* (2), 259-269.

Lears, Jackson T. J. (1985). The concept of cultural hegemony: Problems and possibilities. *The American Historical Review*, *90*(3), 567–593.

Leeds, Eva M., Leeds, Michael A, & Pistolet, Irina. (2007). A stadium by any other name: The value of naming rights. *Journal of Sports Economics*, *8*(6), 581–595.

Light, Duncan, & Young, Craig. (2015). Toponymy as commodity: Exploring the economic dimensions of urban place naming. *International Journal of Urban and Regional Research*, *39*(3), 435–450.

LPF (Liga Profesionista de Fotbal). (2017, February 6). *Despre LPF: Istoric (about LPF: A history)*. Retrieved 22 July 2016 from http://www.lpf.ro/despre-lpf/despre-lpf.html

Massey, Doreen. (1991, June). A global sense of place. *Marxism Today Magazine*, pp. 24–29.

Medway, Dominic, & Warnaby, Gary. (2014). What's in a name? Place branding and toponymic commodification. *Environment and Planning A*, *46*(1), 153–167.

Monitorul. (2016). *Timişoara Townhall [online archives (2012–2016)], Timişoara*. Retrieved from http://www.primariatm.ro/m/

Patton, Michael Quinn (2000). *Qualitative research &evaluation methods* (3rd ed.). London: Sage Publications.

Politehnica, ASU. (2017, March 6). *Istoria clubului (The history of the club)*. Retrieved from http://sspolitehnica.ro/club/politehnica-Timişoara/

Porter, Cristopher. (2015). Loyal to what? FC United's 'shaping walk' through football's 'muck of ages'. *Sports in Society, 18*(4), 452–465.

Putnam, Robert D. (1995). Bowling alone: America's declining social capital. *Journal of Democracy, 6*(1), 65–78.

Rose-Redwood, Reuben. (2011). Rethinking the agenda of political toponymy. *ACME: An International E-Journal for Critical Geographies, 10*(1), 34–41.

Rose-Redwood, Reuben, & Alderman, Derek. (2011). Critical interventions in political toponymy. *ACME: An International E-Journal for Critical Geographies, 10*(1), 1–6.

Rose-Redwood, Reuben, Alderman, Derek, & Azaryahu, Maoz. (2010). Geographies of toponymic inscription: New directions in critical place-name studies. *Progress in Human Geography, 34*(4), 453–470.

Sen, Amartya. (2009). *The idea of justice*. London: Allen Lane.

Simonis, Jana. (2014). *Global-local dialectic*. Retrieved 25 February 2017 from https://centerforinterculturaldialogue.files.wordpress.com/2014/08/key-concept-global-local.pdf

Voiculescu Sorina, Jucu, & Loan Sebastian (2016) Producing urban industrial derelict places: the case of the solventul petrochemical plant in timisoara. *European Urban and Regional Studies, 23*(4), 765-781.

Stoica, Cătălin A. (2004). From good communists to even better capitalists? Entrepreneurial pathways in Post-Socialist Romania. *East European Politics and Societies, 18*(2), 236–277.

Swyngedouw, Erik. (2004). Globalisation or 'glocalisation'? Networks, territories and rescaling. *Cambridge Review of International Affairs, 17*(1), 25–48.

Todorova, Maria, & Gille, Zsuzsa. (2010). *Post-communist nostalgia*. Oxford: Berghahn.

Vesalon, Lucian, & Creţan, Remus. (2013). 'Cyanide kills!': Environmental movements and the construction of environmental risk at rosia montana. *Area, 45*(4), 442-451.

Ward, Kevin. (2008). *Entrepreneurial urbanism*. London: Routledge.

Woisetschläger, David M, Haselhoff, Vanessa J, & Backhaus, Christof. (2014). Fans' resistance to naming right sponsorships: Why stadium names remain the same for fans. *European Journal of Marketing, 48*(7/8), 1487–1510.

Yin, Robert K. (2009). *Case study research* (4th ed.). London: Sage.

Young, Craig, & Light, Duncan. (2001). Place, national identity and post-socialist transformations: An introduction. *Political Geography, 20*(8), 941–955.

Ziesche, Daniel. (2018). 'The East' strikes back. Ultras Dynamo, hyper-stylization, and regimes of truth. *Sport in Society, 21*(6), 883-901.

Data, dispossession, and Facebook: techno-imperialism and toponymy in gentrifying San Francisco

Erin McElroy

ABSTRACT

This article situates the 2015 rebranding of Zuckerberg San Francisco General within San Francisco's technopolitical landscape. After Facebook's CEO Mark Zuckerberg donated to the hospital, thereby acquiring naming rights, local toponymical tension percolated. This was in part due to Facebook's gentrifying role in the city and the racial violence such processes constitute, as well as ongoing practices data colonialism. The latter includes Facebook's attempt at pairing hospital data with user data in order to augment its scope into intimate and bodily geographies. In exploring the intertwining of gentrification and data colonialism, here I forge the concept of techno-imperialism as an analytic and as a point of departure in understanding Facebook's multi-scalar impacts. As I suggest, the transformation of "The General" to "The Zuckerberg" indexes the company's techno-imperiality, as well as its efforts to mask its dispossessive impacts.

A line of seven riot police trample through what appears to be a garden, pushing two people from their home, marked by a long, yellow sign that reads "Foreclosure." If you squint further into the Balmy Alley mural in San Francisco's Mission District, you can see that on one of the officer's helmets is the word "Facebook." Yet that is not the only Facebook reference upon the 2012 "Mission Makeover" mural painted by Lucia Ippolito and her father, Tirso Araiza. An electronic skeletal figure dominating the center of the mural is attempting to plug into a house undergoing eviction, while typing into a white laptop with a Facebook logo on the screen. The electro-skeleton's mouth is fastened shut with a gold zipper, while below, on another panel, youth of color are being handcuffed, aside a protestor holding a protest sign: "Aqui estamos y no nos vamos!" (We're here and we're not going!). As Ippolito describes, the piece portrays two different Missions: one the neighborhood of her youth, and the other, a gentrified landscape indexed by forced displacement, increased racial profiling, the loss of Latinx tenants, and the emergence of a "Luxury Hipster Community for techies with lots of cash" (Gonzalez, 2012). Behold the mutating Mission, where 24th Street meets "Uan Wey" and "Otro Wey" streets. But which way will prevail upon the neighborhood that has undergone the most evictions since the 1990s Dot Com Boom first began to transform the city into what some refer to as the playground for the Silicon Valley elite – a trend only exacerbated during the contemporary era of the Tech Boom 2.0 (McElroy & Szeto, 2018)? Which futures will be foreclosed upon, and which techno-futurist dreams will inhere new forms of gentrification?

As Ippolito and Araiza's mural infers, the gentrification of the Mission is correlative to a real estate marketing strategy that preys upon Mission Latinx culture to boost property value. While this transpired with gusto during the first Dot Com Boom (Graham & Guy, 2002; Mirabal, 2009), it has resurfaced with a vengeance during the contemporary era (Maharawal & McElroy, 2018). For instance, real estate speculator Jennifer Rosdail has rebranded much of the Mission's geography as "The Quad," a quadrilateral "meta hood" in which "quadsters" reside. These new Mission residents, Rosdail describes, "work very hard – mostly in high tech – and make a lot of money," enjoying "the mix of lux and grit" (Rosdail, 2014). By rebranding the historically Latinx neighborhood, one set of cultural geographies is replaced with another. This is premised upon *tech speculation*, or practices in which future real estate value is premised upon the desires of those imbricated in techno-capitalist economies.

The Quad is far from the only rebranding project transpiring in the new Mission. Just blocks away from the Balmy Alley mural lot is the new Zuckerberg General Hospital and Trauma Center. Unlike the formation of the Quad, the renaming what is popularly referred to as "the General" is situated within a different mode of geographic neoliberalism, one linked to a philanthropic model in which private corporations and their CEOs rename public infrastructure – in this case Facebook's Mark Zuckerberg. While literature on *toponymy*, or the geography of place names, has done important work in highlighting the neoliberal violence behind the renaming of urban streets, neighborhoods, and even cities (Berg, 2011; Light & Young, 2015; Madden, 2018; Medway & Warnaby, 2014; Rose-Redwood, 2008), in this paper I question what it means when public health infrastructure is renamed after the world's third wealthiest person who publicly runs, represents, and often stands in for the largest social media platform, one known for questionable data ethics (Metcalf & Carville, 2018; Pew Research Center, 2016). Further, I analyze how the rebranding indexes and intersects with Facebook's gentrifying role in the San Francisco Bay Area. How does the "toponymic cleansing" (Rose-Redwood, Alderman, & Azaryahu, 2010, p. 460) of the General reflect a general trend amongst corporations to mask their malevolent effects, locally and globally, under the liberal auspice of "corporate social responsibility"?

In this article, I position the General's toponymic cleansing as concurrent with a moment in which Facebook has received pushback for its malevolent practices and effects, some of which impact people's health. This occurs locally, in the context of techno-speculative gentrification, and globally, in the context of what Jim Thatcher, David O'Sullivan, and Dillon Mahmoudi refer to as "data colonialism," or utopic promises that facilitate capitalist "accumulation by dispossession" through the colonization and commodification of "everyday life in ways previously impossible" (2016, p. 990). Data colonialism enables what Jeremy Crampton and Andrea Miller define as *algorithmic governance*, or "the manifold ways that algorithms and code/space enable practices of governance that ascribes risk, suspicion and positive value in geographic contexts" (2017, n.p.). While most recently exemplified in the 2018 Cambridge Analytica scandal in which it was revealed that Facebook gave away private user data to third-party retailers and political organizations, likely impacting Donald Trump's electoral success, Facebook's practices of extracting user data for its own benefit run deep. For years the company has employed what Aimee Bahng describes as the "datafication" of human bodies into speculative profit (2015, p. 666).

Thus, one might read the hospital rebranding as a form of *bluewashing*, or corporate practices of masking human and environmental damages (in this case both gentrifying effects and unethical data practices) through corporate social responsibility and the language of liberalism (Berliner & Prakash, 2015; Laufer, 2006). Alternatively, it can be seen as brazen show of force. Or, like the multi-scalar aspect of Facebook's effects and desires themselves, we might see both toponymical impetuses as entangled, as "both-and" (Swanson, Bubandt, & Tsing, 2015, p. 151). In framing this "both-and" approach, here I mobilize the concept of *techno-imperialism*, or the mode in which techno-capitalism expands its power through an array of extractive technologies, penetrating the most intimate and geographically distant spaces alike. As in the case of empires past and present, techno-imperialism accumulates social and economic capital in its imperial hub – a space that from the Cold War through the Dot Com Boom and into the Tech Boom 2.0 has been Silicon Valley and its global metropole of San Francisco (Walker, 2018). Facebook's techno-imperiality relies upon ever-expanding growth models, and bluewashing its less-savory impacts. The toponymics of the Zuckerberg therefore index Facebook's techno-imperialism and its masking.

After a brief overview the rebranding of the hospital and literature on toponymy, in what follows, I study the two scales in which Facebook participates in health ruination, reflected in "The Zuckerberg's" toponymy. First, I examine how its business model constitutes and perpetuates ongoing systems of racialized, gendered, and class-based inequality plaguing the Bay Area. In particular, I assess how these effects impact health in ways that the hospital alone cannot remedy, such as houselessness, mental health crises, and racialized (and at times deadly) algorithmic surveillance. I then turn to Facebook's datafication and algorithmic practices. Both Mark Zuckerberg and his wife, doctor Priscilla Chan, have expressed their century-long goal of eliminating diseases that prey upon human bodies, a goal that the donation reflects, as does their 2016 Chan Zuckerberg Biohub (CZ Biohub) – a biotechnology medical research center in Mission Bay established as part of the Chan-Zuckerberg Initiative (CZI). While their allocating of funds for health betterment can be seen as venerable, simultaneously, Facebook, as a company, has contributed to unethical research and health ruination through unethical data practices. These imperial practices lead to the accumulation of wealth in Silicon Valley and its metropole of San Francisco, thereby enacting a multi-scalar cyclical loop.

The empirics of what follows are largely conceived of in collaboration with the Anti-Eviction Mapping Project (AEMP), a collective that I cofounded in 2013. Producing digital counter-cartographies, data visualizations and analyses, oral histories and videography, community power mapping, and collaborative reports throughout the San Francisco Bay Area region and more recently also in New York City and Los Angeles, the AEMP works with community partners to produce content useful for anti-eviction movement building (Maharawal & McElroy, 2018). All AEMP data referenced in this paper emerges from work that I collaboratively produced with other project members. Additionally, this paper is informed by participant observation conducted while working with the anti-eviction direct action collective, Eviction Free San Francisco (EFSF), which between 2013 and 2016 conducted numerous anti-gentrification direct actions in the Mission District, during the time of the transformation of the hospital. At times we participated in demonstrations, marches, and other forms of protest with nurses and doctors from the General and from the University of California San Francisco (UCSF), and so I also draw upon these conversations and collaborations in the form of participant and non-participant observation. In addition, I read local media on the rebranding of the hospital and resistance to it, as well as on Facebook's forays into data experimentation and datafication. In doing so, I note prevalent themes, also paying attention to the analytics that those representing, critiquing, and protesting institutions hold.

Zuckerberg, toponymy, and the "Ediface Complex"

The rebranding of the General is more than simply a display of a neoliberal urban trend in which "donations" serve as marketing tools. To study the Zuckerberg toponymic transformation is to study the process in which techno-capitalist logics protract into public and bodies alike. As Derek Alderman observes, struggles over place naming are embedded in "larger struggles over social and political identity and are used for resisting the hegemonic order as well as reproducing it" (2000, p. 672). Renaming practices involve not only the addition of a new name but also the deletion of the old, with former names and their constitutive social and political worlds effectively erased as new appellations gain authority. Rueben Rose-Redwood argues that such spatial and intertextual acts of toponymical erasure deeply impact practices of place-making (2008, p. 433). Thus, the hospital rebranding not only marks a semiotic shift upon San Francisco's gentrifying landscapes; the renaming practice itself is agential in neighborhood transformation, in city and regional technopolitics, and in growing a data-driven global empire, all transpiring under the auspices of philanthropic liberalism.

Katherine McKittrick asserts that "naming place is also an act of naming the self and self-histories" (2006, p. xxiii). As she argues, the availability of particular geographies to racialized communities is contingent upon the ability to express and say place. The rebranding of San Francisco's buildings and neighborhoods alike have increasingly foreclosed these possibilities for residents of color, who also disproportionately bear the brunt of evictions (McElroy & Szeto, 2018; Mirabal, 2009). As Lawrence Berg observes, "Neoliberalization is caught up in specific forms of naming that symbolically and materially solidify current (and historical) processes of capitalist accumulation by dispossession" (2011, p. 13). Put otherwise, under the auspices of neoliberalism, renaming practices are not only symbolic; they are speech acts that constitute (and are constituted by) dispossessive materialities. As such, a growing body of urban geographical literature has been revitalized by what Duncan Light and Craig Young describe as a "critical toponymies" approach (2015, p. 435), attentive to the material powers of toponymical inscription upon neoliberalizing landscapes. As they suggest, there is nothing accidental or politically neutral about naming. "Attributing names to places is a way of embedding a particular set of political values into the urban landscape," they write (ibid., p. 436).

And yet, toponymy predates neoliberalism. As a project, it goes back to early colonial histories, which in the Bay Area, began in the eighteenth century when Spanish colonists dispossessed Ohlone communities. Coloniality conferred the status of *terra nullius* (nobody's land) upon California, leaving the Indigenous without land titles. From the naming of San Francisco to the carving out of the Mission, toponymy functioned as a technology of erasure, setting in place settler logics that continue to date. While neoliberal toponymics and gentrification rest upon colonial histories, so does techno-imperialism. As Andrea Miller writes, "Algorithmically driven biometric databases emerge through practices of racialization by the colonial nation-state through modes of data-based extraction and expropriation from the colonized body" (2017, p. 4). This, she marks, is part of "a historical and ongoing arc of protocological racialization that includes eugenics and scientific racism, financialized risk assessment, census-taking and aerial surveying, and predictive policing" (ibid., p. 4). Put otherwise, the data colonization of racialized bodies has long been constitutive of imperialism. While premised upon the accumulation of new forms of increasingly intimate data, and while tethered to racial dispossession through gentrifying effects, techno-imperialism is only made possible through settler logics as they code and recode understandings of private and extractible space and bodies. Thus, I suggest rather than relying upon David Harvey's "accumulation by dispossession" (2004, p. 71) as means to understand neoliberal toponymics, Guarav Jashnani, Priscilla Bustamante, and Brett Stoudt's analytic of "dispossession by accumulation" (2017, p. 19) may be more apropos. This reformulation understands dispossession itself to be accumulative, in this case, the accumulation of settler and neoliberal practices, along with that of data colonization.

As Facebook colonizes data and relies upon settler logics, it accumulates capital that contributes to growing income inequality and gentrification in the Bay Area. In this way, we might read Facebook itself as what Ann Stoler describes as an *imperial formation* (2008), a gradated form of sovereignty, which, unlike empires, is always in the state of becoming. Utilizing technologies defined by "racialized relations of allocations and appropriations," imperial formations "mete out promissory notes that are not exceptions to their operation but constitutive of them" (2008, p. 193). Toponymical bluewashing might be considered one genre of promissory note.

The rebranding of the General took place in 2015, after San Francisco's Board of Supervisors authorized Zuckerberg's $75 million donation. The General is one the region's rare "Level 1" public health institutions, meaning that it attempts to provide high-quality health care to all who need it. Affiliated with UCSF, nurses and doctors often complete residencies in the hospital. Priscilla Chan, Zuckerberg's wife, completed her residency there as a pediatric nurse, in part informing the donation location. The gift transpired soon after the city voted for a $887.4 million bond to expand the General and increase capacity. Not all needed new medical equipment was covered by this, so the $75 million donation offered relief. But it came with strings attached.

It was during EFSF protests and meetings that I heard numerous people describe unease and disgust about the rebranding. Dozens of people promised that they would never utter the new name aloud. Several hospital nurses involved in EFSF reminded us that despite Zuckerberg's donation, voters still paid more than ten times the amount that he contributed. As they asked, why is he more important than "the general" public? Further, even though 83% of San Franciscans voted for the bond measure in 2008, no one was given a chance to vote upon name alteration, which transpired years after. Sasha Cuttler, a nurse at the General, wrote that given that Zuckerberg has accumulated $1.6 billion during the course of one day alone, "if he was working a 12-hour shift like many hospital nurses, then he would earn his contribution in less than an hour" (2015, n.p.). Steve Heilig, director of Public Health and Education for the San Francisco Medical Society, and former worker and volunteer at the General (where he also served on its ethics committee), critiqued: "Asking for, demanding, or being offered 'naming rights'" seems to increasingly *de facto* accompany philanthropic donations, making the donation less of a gift and more of a self-interested promotional strategy (2015, n.p.).

Borrowing from Freud, Heilig diagnosed this contemporary condition as the "Edifice Complex" (though one might call it a "*Ediface* Complex" in thinking through *Face*book's centering of faces and names). Why must a "gift" be accompanied by naming rights, and what materialities and imaginaries does renaming impose upon the gentrifying landscape? Can such a donation really be a considered gift at all? Jacques Derrida suggests that the concept of the gift is often paradoxical as it is nearly impossible to theorize outside of economic exchange logics (1992, p. 7). True gifts, Derrida argues, must not include possibilities of economic returns, nor dialectical relationships between the giver and receiver. Arguably Zuckerberg's donation did not transpire outside of exchange logics and can instead it can be seen as both a show of force and as a means of masking Facebook's effects.

Zuckerberg is not alone in Silicon Valley bluewashing techniques. In 2013, a series of demonstrations broke out against Silicon Valley companies' use of corporate shuttle buses that ferry quadsters from San Francisco to Silicon Valley and back. Property values increase when they can be listed as two blocks from the Google bus stop, or nearby the Facebook bus stop (Goldman, 2013). Further, as the AEMP found, between 2011 and 2014, 69% of "no-fault" evictions issued occurred within four blocks of tech bus stops (Maharawal & McElroy, 2018).[1] Private buses also began delaying public ones, resulting in children missing school breakfasts. In 2014, after tech bus protests received international press, Google donated $7 million to a San Francisco city-wide public bus program that would allow youth free passage. While the program was slated to be passed regardless of Google's donation, Google appeared the hero. Immediately, reporters began contacting EFSF members, asking if protests would subside now that Google was giving back to the community.

While Google's move was not toponymical in nature, other tech companies and CEOs have been more overt. In 2010, Salesforce CEO Marc Benioff donated $100 million the Mission Bay Children's Hospital, now the UCSF Benioff Children's Hospital. Benioff announced the donation during a developer conference at the San Jose McEnery Convention Center, explaining his philanthropic model of giving back. The convention center itself was named after San Jose's former mayor, Tom McEnery, who launched an urban renewal project that displaced a low-income Latinx neighborhood in 1983 (Hukill, 1997). Thus, practices of capital accumulation and dispossession being toponymically masked are nothing new.

Yet toponymic transformation can bear different trajectories and politics. Anti-capitalist organizers, for instance, can also be toponymically honored. In San Francisco, the former Army Street has been changed to César Chávez Street, and Raphael Weill Elementary School is now Rosa Parks Elementary School. In 2017, the Board of Supervisors passed a resolution to transform Justin Herman Plaza to the Embarcadero Plaza until a better name be found. Justin Herman was the 1960s executive director of the San Francisco Redevelopment Agency and responsible for the urban renewal demolition over 1,000 homes and businesses in the largely Black neighborhood of Western Addition. And yet the General remains the Zuckerberg, signaling the techno-imperiality of Facebook, much to the chagrin of those most impacted by gentrification. That said, in 2018, Cuttler proposed renaming the Zuckerberg after José Sarria, a LGBTQ civil rights activist and drag queen who, in 1961, became the US's first openly gay candidate to run for public office. Time will tell if Cuttler's proposal will gain traction.

Tech-speculation, gentrification, and public health

On Ippolito's mural, sitting upon a house facing eviction is a sign indicating a - $51 million sales price. Below it, an advertisement reads, "Call Rich White Realty: www.liveinatrendyhood.com." Carla Wojczuk, a local muralist and AEMP member who leads Mission mural tours told me that when she takes people past Ippolito's mural, they often question the validity of the price tag. She always explains that while the number is speculative, Zuckerberg did purchase a $10 million Mission house in addition to four in Palo Alto for over $30 million (Goldstein, 2016). He was also prepared to purchase 700 acres of land in Hawaii for $100 million which would have denied Indigenous Hawaiians their land rights, but protests and bad press swayed him otherwise (Weise, 2017). The capital used to purchase these properties, like the hospital donation itself, is comprised through an array of techniques, including that of data colonialism. In this way, one form of dispossession incites another. In this section, I assess the gentrifying effects of Facebook's techno-imperialism, particularly as they involve race and biopolitics.

Rich White Realty might only exist in mural form, but from Rosdail to other real estate speculators and corporations such as Zephyr, Vanguard Properties, and Urban Green Investments, housing is created for those with various forms of privilege at the expense of those without, often falling along familiar racial lines. In collaboration with housing clinics such as the Eviction Defense Collaborative (EDC), the AEMP has found that those displaced in San Francisco are disproportionately poor and working-class Black and Latinx residents (Anti-Eviction Mapping Project & Eviction Defense Collaborative, 2017, p. 3) – the inverse demographic privileged in Silicon Valley tech workforces (Maharawal & McElroy, 2018). Facebook, for instance, hires 17% female, 1% Black, and 3% Latinx (Wong, 2017). Seniors and people with disabilities are also disproportionately impacted by evictions (Tenants Together & Anti-Eviction Mapping Project, 2014, p. 8). Not only do Silicon Valley tech companies prefer to hire younger adults (Visier Insights, 2017), but further, private tech shuttle buses such as Facebook's are not disability-friendly, meanwhile posing threats to people in wheelchairs crossing the street in front of them (Vaughan, 2016). As such, it comes as little surprise that resistance to Tech 2.0 gentrification invokes intersectional spatial politics. A 2015 tech bus protest, for instance, was implemented by seniors and people with disabilities, and others have focused on racism and the loss of San Francisco's Black residents.

Gentrification, redevelopment, and public health have long been imbricated in ways that disproportionately impact tenants of color (Dillon, 2018). It was in 2003 that the Planning Department decided to redevelop San Francisco's Trinity Plaza apartments, evicting 360 rent-controlled tenants to construct 1,400 market-rate condominiums (Corburn & Bhatia, 2007, p. 329). Activists rallied against this with support of the Environmental Health Section, the latter highlighting known health impacts of evictions and demolitions (San

Francisco Department of Public Health 2007, pp. 17–19). Since then, anti-displacement organizing has continually invoked the need for an environmental impact review, focused on the public health impacts of gentrification. In 2014, a lawsuit led by the Coalition for Fair Legal and Environmental Transit invoked the need for California Environmental Quality Act review to establish displacement impacts of tech bus stops (Vaughan, 2016). Meanwhile, the ongoing redevelopment of the toxic former shipyards of Bayview-Hunters Point and Treasure Island, much of which is public and Section 8 housing, have led to residents organizing against what Lindsey Dillon writes of as "the health hazards of toxic cleanup and the gentrifying effects of an upscale development project in this historically marginalized area of the city" (2018, p. 227).

Meanwhile, Tech 2.0 correlates with a rise in homelessness, with the Coalition on Homelessness (COH) reporting one in twenty-five public school students unhoused – a figure that doubled between 2011 and 2016 (2016, p. 3). The COH also estimated that 35% of houselessness in San Francisco is related to eviction (2015, p. 16). As it is well established that contexts of eviction and homelessness often precipitate mental health crises (Greenberg & Rosenheck, 2008; Tischler et al., 2007), tech-driven gentrification bears mental health effects. In 2016, Zuckerberg nurses declared a state of emergency when it was revealed that the hospital was turning people away with mental health crises at greater rates than the year prior. Jennifer Friedenbach of the COH, speaking at a nurse-organized rally, described, "Everywhere they're going in the City, they're getting turned away. We're not treating mental health the way we're treating other issues."

In 2014, the AEMP launched an oral history project, Narratives of Displacement and Resistance, chronicling and geolocating personal stories of struggle, protest, and neighborhood change, all embedded in an online and interactive map (Maharawal & McElroy, 2018). Having recorded over 100 stories in San Francisco and Alameda counties as of 2019, with more in the Mission than any other neighborhood, it is nearly impossible to listen and not be confronted with some correlation between the Tech Boom, displacement struggles, and health impacts. For instance, there is the story of Edwin, who after being evicted with his father, was made homeless, while his father, having found refuge in Oakland, became fearful of leaving his new dwelling – a condition that Edwin describes as "root shock." Sometimes root shock can be deadly. For instance, in an AEMP collaboration with the EDC documenting where 500 people evicted in 2012 ended up post-eviction, we found that fourteen people ended up houseless, and that two passed away (Anti-Eviction Mapping Project & Eviction Defense Collaborative, 2017). In 2017, Iris Canada, a 100-year-old Black resident-turned-housing-activist was evicted from her home of 60 years. She died shortly after. In 2013, Mary Elizabeth Phillips received an eviction notice at the age of 97 on the edge of the Mission, and although she fought back for the ability to stay in her home, her neighbor and caretaker was evicted, endangering her care. After Mary passed away at the age of 100, her building was flipped to house wealthy tech workers. In 2015, Fergus O'Sullivan, known for his large eviction portfolio and utilization of Airbnb post-eviction, displaced two 95-year-old Latina women from their home on York Street just one block from the General. One of them died two weeks later. Years before, in 2012, O'Sullivan also evicted Michael, who was interviewed for the AEMP's oral history project. At the time of the interview, Michael, who had disabling HIV/AIDS, was houseless. His former unit is now a short-term rental.

In his interview about his father's root shock, Edwin also describes signs that his neighborhood was transforming. Prior to his eviction, new neighbors racially profiled him and his father for taking out the garbage in their own house, reporting them to the police. Around the same time, Alex Nieto, whose parents we also interviewed for the oral history project, was murdered by four Mission Station San Francisco Police Department officers after being racially profiled by a series of young White men on Bernal Hill, all with tech jobs and new to the neighborhood.

Not only was Alex's death embedded in the longstanding US tradition of racialized police violence, but further, he had been known to have mental health struggles, as many victims of police violence do. Such forms of policing have been, in recent years, described as part of a public health crisis (Cloud & Davis, 2015). Community activists also linked Alex's death to gentrification. UCSF physician and founder of the Do No Harm Coalition, Rupa Marya, reflected, "With the recent killings of young black and brown people in San Francisco by the police and the ongoing displacement of people for profit, a gaping wound has opened in this body [politic] and we can see that this wound is festering" (2016, n.p.). Meanwhile, collectives such as the Anti-Police Terrorism Project responded with calls for "the immediate divestment of city funds for policing and investment in sustainable, affordable housing so Black, Brown and Indigenous people can remain in their hometowns of Oakland and San Francisco" (2015, n.p.). Calls such as these accord with scholarship that understands a correlation between racialized policing, public health crises, and gentrification (Camp, 2012; Maharawal, 2017). After community demands to end police impunity were unanswered, members of the collective, along with other community groups, held a performative people's trial outside of the Mission District police station, which included the blockading of a Facebook bus, contesting linkages of tech, eviction, houselessness, and new forms of racialized surveillance and policing.

It was not lost on protesters blockading the Facebook bus – a blue iteration of what Rebecca Solnit has described as "alien overlords" ruling over San Francisco (2013) – that Facebook's facial recognition algorithms are agential in comprising the world's largest biometric database, one with 2.2 billion users (Dishaw, 2015). The database is used by local police officers upon their discretion, particularly to track protestors, including Black Lives Matter protestors fighting racialized policing (Coleman, 2016; Guynn, 2016; Joseph, 2017; Lomas, 2017). In 2018, Amazon began providing law enforcement with its Rekognition technology to track and identify "people of interest," which, as the American Civil Liberty Union (ACLU) noted, includes undocumented immigrants and Black activists (Cagle & Ozer, 2018, n.p.). Facebook has estimated that it can accurately identify a tagged person 97% of the time through its DeepFace facial recognition algorithm, better than the FBI's own Generation Identification software, which only boasts a 50% to 85% accuracy rating (LaChance, 2016). The human brain is apparently only .28% better at recognizing people than Facebook's nine-layer neural network that matches faces with names (Campione, 2018). In 2017, Facebook went a step further, acquiring a new patent to track users' emotions from their phone and computer cameras in real time as they scroll through their Facebook feed (CBInsights, 2017). Meanwhile, the company has spent billions of dollars buying up potential competitors such as Face.com, Masquerade, and Faciometrics (Campione, 2018). In this way Facebook, along with other providers of biometric surveillance technology, garner techno-imperial monopolies on algorithmic governance infrastructure.

Algorithmic practices often replace human decision-making processes with machine-based learning, feigning objectivity. And yet, algorithms have been constituted through human-made fantasies and prejudices, coded and spatialized to produce desired results, protocols, predictions, and borders through practices such as predictive policing (Amoore, 2013). Predictive policing entails the use of algorithmic systems to determine "hot spots" and people of interest, often using race as a surveillance indicator. In profiling neighborhoods and streets, predictive policing constitutes what Emily Kaufman describes as "bio-spatial" profiling, in which militarized urbanism and advanced policing technologies rely upon the racial demarcation of both people and spaces (2016, p. 74). This mapping paves the way for gentrification, cleansing neighborhoods of those that inhibit revalorization processes (2016, p. 80). And yet, predictive policing companies such as the Santa Cruz-based startup PredPol (whose chairman of the board is Facebook's former Chief Privacy Officer and Head of Global Public Policy), claim that because their data is algorithmically derived, it is less racist than analog human judgment (Robinson & Koepke, 2016).

While San Francisco is not yet using the same level of predictive policing as other cities, it had extensive talks with PredPol in 2012 and 2013, and it is unknown if a deal will be eventually cemented (Robinson & Koepke, 2016, p. 17). In her 2018 electoral campaign, San Francisco mayor London Breed praised predictive policing, highlighting the work that she had already conducted in Taraval and the Mission to generate a data-driven city response network to monitor and track serial offenders (Breed, 2018). Meanwhile, San Francisco's metro system, Bay Area Rapid Transit, has also been using predictive policing to surveil, bait, and catch bike thieves (CBS SF, 2015). While it is unknown if Facebook is currently working with local police, there always lingers the possibility that, as Marc Rotenberg of the Electronic Privacy Information Center worries, "Facebook might sell the information to retailers or be forced to turn it over to law enforcement – in both cases without users knowing it" (Rosenblatt, 2016, n.p.). As has been uncovered through ACLU record requests, of 63 police departments interrogated in California, twenty, including Oakland's and San Jose's, have admitted to having acquired social media surveillance software to extract data from Facebook, Instagram, and Twitter (Wong, 2016). In this way, Facebook technology entwines with contexts of racialized surveillance, a process also entangled in gentrification processes.

Techno-imperialism and data dissection

Not only might people not know if their personal data is being used by police in instances of algorithmic governance, but also, many people are unaware that their unpaid labor of uploading, geolocating, and tagging photographs on Facebook or even Instagram (owned by Facebook) feeds surveillance regimes. Even if someone without a Facebook account visits a non-Facebook webpage and clicks that they "like" a certain article, Facebook keeps track of that information through a third-party request (Gillmor, 2018). Thus, Facebook functions by exploiting users unpaid labor, selling data cleaned in this unpaid labor to other companies. Neda Atanasoski and Kalindi Vora suggest that such practices transpose "categories of abjection from the human into the very infrastructure upon which the techno-utopic future depends" (2015, p. 25).

Not only is abjection created through techno labor practices, but also humans whose data are dissected and parsed by algorithmic technologies are also rendered less than human. Building upon this, in what follows, I turn from techno-entanglements of algorithmic governance, gentrification, and policing to those of data colonization and public health. Specifically, I look at Facebook's history of nonconsensual data usage, which includes health data and which impacts health. As I argue, datafying users into profitable parts while dissecting and compiling personal information both constitutes and indexes the company's techno-imperiality.

Digital lives, as inextricably entangled in non-digital ones, have become fodder for capitalist accumulation (Boelstorff, 2016). Although Facebook is informed by (and informs) the material world, Facebook has decided that digital avatars and their interfaces are undeserving bodies and spaces. And yet, Facebook itself has conducted studies proving inextricable connectivity between digital and non-digital lives. In 2012, it manipulated content on 689,003 users to decipher effects of "emotional contagion." With researchers from Cornell University and UCSF, the company altered modes of positive and negative content on users' pages, leading to the conclusion that "emotions expressed by others on Facebook influence our own emotions, constituting experimental evidence for massive-scale contagion via social networks" (Kramer, Guillory, & Hancock, 2014, p. 8788). In 2014, after researchers published their results, internet activists, academics, and politicians expressed outrage, calling the psychological manipulation "spooky," "creepy," "scandalous," "invasive," and "disturbing" (Corbett, 2014, n.p.; LaFrance, 2014, n.p.). Jim Sheridan from the British Commons media select committee described, "If people are being thought-controlled in this kind of way there needs to be protection and they at least need to know about it" (Booth, 2014, n.p.). While Facebook claimed that the experiment was "consistent with Facebook's data use policy, to which all users agree prior to creating an account on Facebook, constituting informed consent for this research," others, including Princeton psychology professor Susan Fiske who edited the findings, agree that in effect, the experiment was nonconsensual (LaFrance, 2014, n.p.). Others were disturbed that the study had been approved by an institutional review board "on the grounds that Facebook filters user news feeds all the time, per the agreement" (Meyer, 2014, n.p.).

While Facebook eventually apologized, at first it defended its methods, claiming that the study was necessary for the company's development of new features, and consistent with their culture of innovation (Gibbs, 2014). Years before, in 2008, Illinois residents sued Facebook for stealing the property of their digital identities, causing them real harm and non consensual "informational injury" (Rosenblatt, 2016, n.p.). As the plaintiffs argued, "Just as trade secrets or subscriber lists can be proprietary to a company like Facebook, unique and unchangeable biometric identifiers are proprietary to individuals." Facebook transferred the case, *Facebook Biometric Information Privacy Litigation 15-cv-03747*, to San Francisco, where it has since twice argued for dismissal on grounds that no concrete injury can be proven. Thus, while capital both absorbs and is constituted by digital and analog data-mining, non-consensually dismembering, analyzing, and selling fragments of bodies and lives, when it comes to litigation, only the analog is recognized.

However, ten years later, in March 2018, a new story materialized that Facebook could no longer repress or belittle. As was revealed, a British stealth political technology firm, Cambridge Analytica, had worked with Facebook to harvest user data, mobilizing information used to benefit Donald Trump's 2016 US electoral victory. An undercover investigation by the UK's Channel 4 showed that Analytica was partly owned by rightwing propagandist and Trump appointee Steve Bannon, but moreover, Analytica possessed 230 million American voters' profiles based upon 5,000 detailed data points largely and consensually obtained from Facebook (Kroll, 2018). While social media datamining well predates Analytica, the company had developed unique micro-targeting techniques to pair consumer information with psychological data extracted from Facebook. This was used by the Trump campaign to target people with custom images, messages, ads, mailers, and even in-person interactions – a recipe that the company's CEO, Alexander Nix, called its"secret sauce" (Kroll, 2018, n.p.). The scandal further revealed Facebook's collusion with hedge fund investor Robert Mercer to target swing states with Islamophobic messaging during the election. In 2018, it was also revealed that Black Americans on Facebook and Instagram were uniquely targeted during the 2016 elections to favor Trump (Danielle, 2018).

Facebook admitted their error after the scandal surfaced, apologizing that any of their users could have had their data stolen by "malicious actors" (Cara, 2018, n.p.). In response, in early April 2018, the company introduced new data-sharing restrictions. Per the company's Chief Operating Officer Sheryl Sandberg, "What we didn't do until recently and what we are doing now is just take a broader view looking to be more restrictive in ways data could be misused" (Meyer, 2018, n.p.). However, not only had Facebook already been long experimenting with user data, but it had also been selling user data to third parties for years. The company implemented its ad platform in 2007; today, one in five dollars spent on online advertising goes to Facebook (Collins & Buchanan, 2018). In fact, as soon as implementing the "like" button in 2009, Facebook began providing information to advertisers. In 2012, Custom Audiences was introduced, allowing merchants to select their targets based on demographics (ibid., 2018). In 2014, Lookalike Audiences and Ethnic Affinity were realized, allowing retailers and organizations to target based upon user similarities, including what Facebook determines a users' race or ethnicity to be (even though users cannot themselves select a race or ethnicity). As such, advertisers can now algorithmically target: "Anyone who lives in Philadelphia, studies philosophy in college, is 21, has bought a blue T-shirt in the past year, is neurotic, makes less than $28,000 a year, is likely to buy a minivan in the next six months, is interested in camping and whose interests align with those of African-Americans. Plus, anyone on Facebook who is similar to them" (Collins & Buchanan, 2018, n.p.).

While the Analytica scandal galvanized international attention and resentment, inciting the boycott campaign #DeleteFacebook (and then later a call by the National Association for the Advancement of Colored People to #LogOutFacebook), less illuminated was another privacy breach involving hospital data. As CNBC uncovered, Facebook's secretive research arm, "Building 8," was in the midst of launching new experiments by extracting user information to pair with hospital data. In March 2018, Building 8 sent a doctor to talk to top US hospitals, including the American College of Cardiology (ACC) and the Stanford University School of Medicine. The idea, according to Facebook, was for hospitals to share anonymized medical information such as health issue and age to match it with clients' Facebook accounts. This would allow Facebook to offer information to hospitals that might help improve treatment. As Facebook told the news agency:

> The medical industry has long understood that there are general health benefits to having a close-knit circle of family and friends. But deeper research into this link is needed to help

> medical professionals develop specific treatment and intervention plans that take social connection into account. With this in mind, last year Facebook began discussions with leading medical institutions, including the American College of Cardiology and the Stanford University School of Medicine, to explore whether scientific research using anonymized Facebook data could help the medical community advance our understanding in this area (Farr, 2018, n.p.).

Facebook gave the example of an elderly user without many local friends, who might benefit from post-surgery visits. According the Facebook, data sharing would help compare "what a health system knows about its patients (such as: person has heart disease, his age 50, takes 2 medications and made 3 trips to the hospital this year) with what Facebook knows (such as: user is age 50, married with 3 kids, English is not a primary language, actively engages with the community by sending a lot of messages)" (Farr, 2018, n.p.). And yet, Facebook promised that its data would be anonymized. A Facebook spokesperson contextualized, "The project would not attempt to provide health recommendations for specific people. Instead the focus would be on producing general insights that would help medical professionals take social connectedness into account as they develop treatment or intervention programs for their patients" (Kastrenakes, 2018, n.p.). Data accesses would be limited to selected Facebook employees and medical research partners at the hospitals. Facebook also provided a statement by ACC's interim CEO, who confirmed that the project would use anonymized social media data to help prevent and treat heart disease, the world's leading cause of death. Facebook proposed using a cryptographic technique known as *hashing*, in which its data could be matched with hospitals' while obscuring identifiable information. However, this pseudonymization technique is not in fact anonymization, as data is thus left open to re-identification.

This foray into health data occurred concurrently with other Facebook medical industry ventures. In New York, it has been growing a "Facebook health" team, aimed to help pharmaceutical companies target users with particular demographic profiles or who have "liked" health advocacy pages (Farr, 2018). Meanwhile, the CZI announced an ambitious aim of curing all diseases during the lifetime of today's children. Invoking Google's initiative to "cure death," the philanthropic CZI will see $3 billion invested over the next ten years, and $45 billion over the course of Chan and Zuckerberg's lifetimes, aimed to "alleviate suffering" (Alba, 2016). CZI donated $600 million to the CZ Biohub with this in mind, aiming to bolster communication between researchers in local universities while developing new scientific tools, artificial intelligence, and machinery. While this may appear an honorable ambition, it is hard to separate it from Facebook's own techno-imperial endeavors, for instance, new semi-secretive hardware projects emerging from Building 8, which include brain scanning technology being led by a neuroscientist who helped develop a mind-controlled prosthetic arm (Heath, 2017). How divorced from its own proprietary advances will Facebook be from emergent work funded through the CZI, and what comes first – capitalist desires or those of disease prevention? Further, as members of the Science Feminist Anti-Racist Equity (FARE) collective critiqued, much needs to change to so that Biohub's research ensures treatment affordability and access, irrespective of race, class, gender, or disability, and livable wages for contract employees (2016). As they also question, how the Biohub will prioritize the health of patients over biotech patents, and how it will attend to the "upstream" causes of disease, such as lack of affordable housing, evictions, homelessness, and criminalization (2016, p. 371).

Contexts such as these, along with the data breaches revealed by Cambridge Analytica, spawned new anxieties amongst nurses and clients at the newly named

Zuckerberg Hospital. While the impact of the scandal led Facebook to press pause on their hospital data project (Farr, 2018), public trust was already compromised, particularly in the Zuckerberg. Heather Ali, a hospital employee, reported to the *New York Times*, "We are in charge of keeping our most vulnerable people private and protected. Now people wonder, 'How much is my privacy protected at a hospital with that name on it?'" (Bowles & Kang, 2018, n.p.). Another nurse, Megan Brizzolara, reported that the name Zuckerberg "scares" patients (Collins & Buchanan, 2018, n.p.). Nurse Sasha Cuttler, concurred, explaining that "People are afraid. I've spoken with people who have said, 'I'm afraid to tell my doctor anything, because I don't know who is going to get that information'" (Ward, 2018, n.p.).

Irrespective of Facebook, patient data had already been compromised at the General/ Zuckerberg. In 2017, a former employee of Nuance Communications illegally accessed nearly 900 detailed records for medical transcription work. In 2014, over 56,000 medical hospital records were stolen from the billing company with whom the hospital contracts (Ho, 2018). Further, as one nurse-in-training who works for the San Francisco's Department of Public Health told me after an EFSF meeting, in order to receive services, houseless people are required to sign forms giving away their personal data to the city, and really have no way of opting out of doing so if they need care. Thus, houseless people, many of whom lack houses due to gentrification, are forced to concede their data to the city to receive care.

Conditions such as these, along with growing wariness of Facebook's data ethics, incited a second nurse-led protest against Zuckerberg toponymics. Like their first demonstration after the name change, nurses brought together concerns of datafication and gentrification. "It's fine to have somebody's name and to accept a donation and fundraising, but that doesn't mean you can do whatever you like with patient data," Cuttler contextualized (Ward, 2018, n.p.). Nevertheless, the name change itself was still upsetting to him and other nurses, who saw it as part of tech's broader urban invasion. Another nurse, Guy Vandenberg, wore scrubs and a placard that declared the diagnosis of "Malwareberg." The prescription was also noted: "Uninstall Zuckerberg." Meanwhile, then interim Mayor Mark Farrell, known to side with big money at the expense of those most impacted by gentrification, commented on the local CBS news station that the naming of the Zuckerberg and issues of patient data confidentiality should not be seen as connected (Ward, 2018). Former supervisor and anti-gentrification advocate John Avalos, who voted in favor of Zuckerberg's donation, now has regrets: "Had we known what we know now, perhaps we wouldn't have accepted the funds from Zuckerberg" (Bowles & Kang, 2018, n.p.).

If Zuckerberg, Facebook, or the Biohub really endeavored to eliminate diseases as they claim, one could argue that a first step might be the cessation of user data mining. Another might be gentrification mitigation, what the FARE collective describes as gentrification's upstream causes. While the CZI did donate $3.6 million toward housing crisis relief in 2017, mostly concentrated south of San Francisco in San Mateo County, it is hardly enough to mitigate impact. Further, as housing organizers learned in a regional organizing meeting, during this time a Facebook employee had been hoping to donate anonymously, until Facebook Corporate got wind and insisted the donation be done in their name. Debate ensued amongst organizers as whether to take the donation, as it could go a long way in supporting rent control campaigns, but at the same time, crediting Facebook undermined the deleterious impacts of the corporation. As some activists interrogated, why could the company not simply donate anonymously?

Afterlives

"Death is the final frontier of cyberspace," writes Daniel Greenwood, reviewing John Romano and Evan Carroll's 2010 *Your Digital Afterlife: When Facebook, Flickr and Twitter Are Your Estate, What's Your Legacy?* (Digital Afterlife, 2017, n.p.). By this, he conceptualizes the new terrain of postmortem cyber identities, such as Facebook profiles of people no longer living. While Greenwood is excited by afterlife temporalities, the material conditions for this posthuman spatiotemporal fantasy rely upon actual deaths. This does not imply that the virtual is unreal, but rather that Facebook's "digital, non-digital, and not-only-digital" materialities are grounded in extractive technologies both accumulating and derived in the Bay Area (Crampton & Miller, 2017, n.p.). Facebook's algorithmic laboratory is of course global, telescoping techno-imperialism far and wide (LaFrance, 2016). But it also concentrates locally, supported by a workforce gentrifying Quad-like geographies, where Facebook employees pay exorbitant rent checks derived from big data labor. Real estate speculators capitalize upon this, displacing tenants, contributing to conditions of death and disease. These lives and deaths are thus the other afterlives constitutive of cyberspace's final frontier. But so are the bodies of hospital patients, whose information may soon be paired with Facebook user information. How different might these bodies and their digital dissections be from those already surveilled by DeepFace and other technologies of algorithmic governance and datafication? How does one reconcile these futures, while mourning those already disappeared through processes of racialized surveillance, gentrification, and data colonization?

By toponymically erasing "the general public" from the hospital, both the lives and deaths of the general public (digital and otherwise) become further dispossessed from edifices of public memorialization. As such, the renaming of the General can be understood as a form of toponymic cleansing, in which the remembrance of some (or in this case one) depends upon the erasure of others (in this case, many). Thereby, in the words of Light and Young, "naming is one element of strategies among elites to construct and institutionalize a hegemonic narrative of history" (2015, p. 437). By privatizing the naming rights of the General, Zuckerberg and Facebook become official writers of San Francisco history and memory, literally inscribing the walls of the city's public hospital with a narrative structure in which Facebook is salvific and good. In this narrative, the lives and deaths disrupted by techno-imperialism are rendered nameless. The naming of the Zuckerberg is thus part of a toponymical algorithm, concordant with "the political aspirations and agenda of the incoming regime" (2015, p. 437). Zuckerberg's gift deletes histories antithetical to the company's techno-imperial vision, while simultaneously marking its imperiality. Yet on other edifices, not far away, murals such as Ippolito's hold memories of other histories, futures, and names.

Note

1. Evictions in San Francisco are categorized as either "fault" or "no-fault." Fault evictions imply that the landlord evicts due to some breach of lease, whereas no-fault imply that the tenant did not violate their lease and are nevertheless evicted.

Disclosure statement

No potential conflict of interest was reported by the author.

ORCID

Erin McElroy http://orcid.org/0000-0002-8130-9055

References

Alba, Davey. (2016, September 21). Zuckerberg and Chan promise $3 billion to cure every disease. *Wired*, Retrieved from https://www.wired.com/2016/09/zuckerberg-chan-promise-3-billion-curing-disease

Alderman, Derek. (2000). A street fit for a king: Naming places and commemoration in the American South. *Professional Geographer*, *52*(4), 672–684.

Amoore, Louise. (2013). *The politics of possibility: Risk and security beyond probability*. Durham: Duke University Press.

Anti-Eviction Mapping Project & Eviction Defense Collaborative. (2017). Eviction report (2016). Retrieved from https://antievictionmap.com/s/EDC_2016.pdf

Anti-Police Terror Project. (2016, August). New blog. Retrieved from http://www.antipoliceterrorproject.org/new-blog/

Atanasoski, Neda, & Vora, Kalindi. (2015). Surrogate humanity: Posthuman networks and the (racialized) obsolescence of labor. *Catalyst: Feminism, Theory, Technoscience*, *1*(1), 1–40.

Bahng, Aimee. (2015). Specters of the Pacific: Salt fish drag and atomic hauntologies in the era of genetic modification. *Journal of American Studies*, *49*(4), 663–683.

Berg, Lawrence. (2011). Banal naming, neoliberalism, and landscapes of dispossession. *ACME: an International E-Journal for Critical Geographies*, *10*(1), 13–22.

Berliner, Daniel, & Prakash, Aseem. (2015). "Bluewashing" the firm? Voluntary regulations, program design, and member compliance with the United Nations Global Compact. *Policy Studies Journal*, *43*(1), 115–138.

Boellstorff, Tom. (2016). For whom the ontology turns: Theorizing the digital real. *Cultural Anthropology*, *57*(4), 387–407.

Booth, Robert. (2014, June 30). Facebook reveals news feed experiment to control emotions. *The Guardian*, Retrieved from: https://www.theguardian.com/technology/2014/jun/29/facebook-users-emotions-news-feeds

Bowles, Nellie, & Kang, Inyoung. (2018, May 15). San Francisco nurses protest the Zuckerberg hospital's name. *New York Times*, Retrieved from https://www.nytimes.com/2018/05/15/us/california-today-san-francisco-zuckerberg-hospital.html

Breed, London. (2018, February 15). Making a safer San Francisco. *Medium*, Retrieved from https://medium.com/@LondonBreed/making-a-safer-san-francisco-108b002ebc91

Cagle, Matt, & Ozer, Nicole. (2018, May 22). Amazon teams up with government to deploy dangerous new facial recognition technology. *ACLU Blog*, Retrieved from https://www.aclu.org/blog/privacy-technology/surveillance-technologies/amazon-teams-government-deploy-dangerous-new

Camp, Jordan. (2012). Blues geographies and the security turn: Interpreting the housing crisis in Los Angeles. *American Quarterly*, *64*(3), 653–678.

Campione, Riccardo. (2018, March 30). What entrepreneurs need to know about facial recognition technology. *Entrepreneur*, Retrieved from https://www.entrepreneur.com/article/311228
CBInsights. (2017, June 1). Facebook's emotion tech: Patents show new ways for detecting and responding to users' feelings. Retrieved from https://www.cbinsights.com/research/facebook-emotion-patents-analysis
CBS. (2015, January 7). BART using predictive policing to nab bike thieves as thefts spike 21 percent systemwide. Retrieved from https://sanfrancisco.cbslocal.com/2015/01/07/bart-using-predictive-policing-to-nab-bike-thieves-with-thefts-up-21-percent-systemwide
Cloud, David, & Davis, Chelsea. (2015). *First do no harm: Advancing public health in policing practices*. New York, NY: Vera Institute of Justice.
Coalition on Homelessness. (2016). Beating the odds (Annual report, 2016). Retrieved from http://www.cohsf.org/wp-content/uploads/2014/08/2016-ANNUAL-REPORT-COH-1.pdf
Coleman, Mat. (2016). State power in blue. *Political Geography*, *51*(1), 76–86.
Collins, Keith, & Buchanan, Larry. (2018, April 11). How Facebook lets brands and politicians target you. *New York Times*, Retrieved from https://www.nytimes.com/interactive/2018/04/11/technology/facebook-sells-ads-life-details.html
Corbett, Jessica. (2014, July 3). New questions, few answers in Cornell's Facebook experiment. *The Ithaca Voice*, Retrieved from https://ithacavoice.com/2014/07/new-questions-answers-cornells-facebook-experiment
Corburn, Jason, & Bhatia, Rajiv. (2007). Health impact assessment in San Francisco: Incorporating the social determinants of health into environmental planning. *Journal of Environmental Planning and Management*, *50*(3), 323–341.
Crampton, Jeremy, & Miller, Andrea. (2017, May 9). Algorithmic governance. *Antipode Blog*, Retrieved from https://antipodefoundation.org/2017/05/19/algorithmic-governance/
Cuttler, Sasha. (2015, December 20). San Francisco's General Hospital belongs to the public – It's not Zuckerberg's. *Alternet*, Retrieved from http://www.alternet.org/personal-health/san-franciscos-general-hospital-belongs-public-its-not-zuckerbergs-plaything
Danielle, Britni. (2018, December 19). Here's why the NAACP wants you to boycott Facebook. *Essence*. Retrieved from https://www.essence.com/news/naacp-facebook-boycott/
Derrida, Jacque. (1992). *Given time: Counterfeit money*. (Peggy Kamuf, Trans.). Chicago: University of Chicago Press.
Digital Afterlife. (2017). Retrieved from http://www.yourdigitalafterlife.com.
Dillon, Lindsey. (2018). The breathers of Bayview Hill: Redevelopment and environmental justice in Southeast San Francisco. *Hastings Environmental Law Journal*, *24*(2), 227–236.
Dishaw, Thomas. (2015, October 26). Facebook's person-centric biometric database is bigger than the FBI's. *Investment Watch Blog*, Retrieved from https://investmentwatchblog.com/facebooks-person-centric-biometric-database-is-bigger-than-the-fbis
Farr, Christina. (2018, April 5). Facebook sent a doctor on a secret mission to ask hospitals to share patient data. *CNBC*, Retreived from https://www.cnbc.com/2018/04/05/facebook-building-8-explored-data-sharing-agreement-with-hospitals.html
Gibbs, Samuel. (2014, July 3). Facebook policy head says emotion experiments were "innovative." *The Guardian*, Retrieved from https://www.theguardian.com/technology/2014/jul/03/facebook-emotion-experiments-monika-bickert
Gillmor, Daniel. (2018, April 5). Facebook is tracking me even though I'm not on Facebook. *ACLU Blog*, Retrieved from https://www.aclu.org/blog/privacy-technology/internet-privacy/facebook-tracking-me-even-though-im-not-facebook
Goldman, Alexandra. (2013). *The 'Google shuttle effect:' Gentrification and San Francisco's Dot Com Boom 2.0* (Unpublished master's thesis). University of California, Berkeley, CA.
Goldstein, Daniel. (2016, May 14). Facebook CEO Mark Zuckerberg to raze four homes with views of his house. *MarketWatch*, Retrieved from http://www.marketwatch.com/story/facebook-ceo-mark-zuckerberg-to-raze-four-homes-with-a-view-of-his-own-2016-05-24
Gonzalez, Lucia. (2012). Mission Makeover mural on Kickstarter. Retrieved from https://www.youtube.com/watch?v=_AN-YHs0Zxc

Graham, Steven, & Guy, Simon. (2002). Digital space meets urban place: Sociotechnologies of urban restructuring in downtown San Francisco. *City*, *6*(3), 369–382.

Greenberg, Greg, & Rosenheck, Robert. (2008). Jail incarceration, homelessness, and mental health: A national study. *Psychiatric Services*, *59*(2), 170–177.

Guynn, Jessica. (2016, October 11) ACLU: Police used Twitter, Facebook to track protests. *USA Today*, Retrieved from https://www.usatoday.com/story/tech/news/2016/10/11/aclu-police-used-twitter-facebook-data-track-protesters-baltimore-ferguson/91897034/

Harvey, David. (2004). The "new imperialism": Accumulation by dispossession. *Actuel Marx*, *1*, 71–90.

Heath, Alex. (2017, April 17). Facebook is about to take the wraps off its secretive and ambitious consumer hardware group. *Business Insider*, Retrieved from https://www.businessinsider.com/facebooks-building-8-working-on-camera-augmented-reality-mind-reading-projects-2017-3/

Heilig, Steven. (2015, April 2). Please don't change San Francisco General's name. *San Francisco Chronicle*, Retrieved from http://www.sfchronicle.com/opinion/openforum/article/Please-don-t-change-San-Francisco-General-s-6176553.php.

Ho, Catherine. (2018, May 11). Data breach affects nearly 900 patients from two San Francisco hospitals. *SF Chronicle*, Retrieved from https://www.sfchronicle.com/business/article/Data-breach-affects-900-patients-from-two-San-12908583.php

Hukill, Traci. (1997, September 25). Your name in lights. *Metro Active*, Retrieved from http://www.metroactive.com/papers/metro/09.25.97/cover/naming1-9739.html

Insights, Visier. (2017, September). The truth about ageism in the tech industry. Retrieved from https://www.visier.com/wp-content/uploads/2017/09/Visier-Insights-AgeismInTech-Sept2017.pdf

Jashnani, Gaurav, Bustamante, Prisilla, & Stoudt, Brett. (2017). Dispossession by accumulation: The impacts of discretionary arrests in New York City. *Race and Justice*, 1–28. doi:10.1177/2153368717742476

Joseph, George. (2017, February 8). Cellphone spy tools have flooded local police departments. *City Lab*, Retrieved from https://www.citylab.com/equity/2017/02/cellphone-spy-tools-have-flooded-local-police-departments/512543/

Kastrenakes, Jacob. (2018, April 5). Facebook spoke with hospitals about matching health data to anonymized profiles. *The Verge*, Retrieved from https://www.theverge.com/2018/4/5/17203262/facebook-medical-data-sharing-plan-healthcare

Kaufman, Emily. (2016). Policing mobilities through bio-spatial profiling in New York City. *Political Geography*, *55*(1), 72–81.

Kramer, Adam, Guillory, Jaimie, & Hancock, Jeffrey. (2014). Experimental evidence of massive-scale emotional contagion through social networks. *Pnas*, *111*(24), 8788–8790.

Kroll, Andy. (2018, May/June). Cloak and data. The real story behind Cambridge Analytica's rise and fall. Mother Jones, Retrieved from https://www.motherjones.com/politics/2018/03/cloak-and-data-cambridge-analytica-robert-mercer/

LaChance, Naomi. (2016, May 18). Facebook's facial recognition software is different from the FBI's: Here's why. *NPR: All Tech Considered*, Retrieved from http://www.npr.org/sections/alltechconsidered/2016/05/18/477819617/facebooks-facial-recognition-software-is-different-from-the-fbis-heres-why

LaFrance, Adrienne. (2014, June 28). Even the editor of Facebook's mood study thought it was creepy *The Atlantic*, Retrieved from https://www.theatlantic.com/technology/archive/2014/06/even-the-editor-of-facebooks-mood-study-thought-it-was-creepy/373649/

LaFrance, Adrienne. (2016, February 11). Facebook and the new colonialism. *The Atlantic*, Retrieved from https://www.theatlantic.com/technology/archive/2016/02/facebook-and-the-new-colonialism/462393/

Laufer, William. (2006). *Corporate bodies and guilty minds: The failure of corporate criminal liability*. Chicago: University of Chicago Press.

Light, Duncan, & Young, Craig. (2015). Toponymy as commodity: Exploring the economic dimensions of urban place names.". *International Journal of Urban and Regional Research*, *39*(3), 435–450.

Lomas, Natasha. (2017, February 13). ACLU calls for tech firms to lobby for surveillance reform. *Tech Crunch*, https://techcrunch.com/2017/02/13/aclu-calls-for-tech-firms-to-lobby-for-surveillance-reform/

Madden, David. (2018). Pushed off the map: Toponymy and the politics of place in New York City. *Urban Studies, 55*(8), 1599–1614.

Maharawal, Manissa. (2017). Black Lives Matter, gentrification and the security state in the San Francisco Bay Area. *Anthropological Theory, 17*(3), 338–364.

Maharawal, Manissa, & McElroy, Erin. (2018). The Anti-Eviction Mapping Project: Counter mapping and oral history towards Bay Area housing justice. *Annals of the American Association of Geographers, 108*(2), 380–389.

Marya, Rupa. (2016, July 29). The open, festering wound of injustice. *48 Hills*, Retrieved from https://48hills.org/2016/07/the-open-festering-wound-of-injustice/

McElroy, Erin, & Szeto, Andew. (2018). The racial contours of YIMBY/NIMBY Bay Area Gentrification. *Berkeley Planning Journal, 29*(1), 7–44.

McKittrick, Katherine. (2006). *Demonic grounds: Black women and the cartographies of struggle.* Minneapolis: University of Minnesota Press.

Medway, Dominic, & Warnaby, Gary. (2014). What's in a name? Place branding and toponymic commodification. *Environment and Planning A, 46*(1), 153–167.

Metcalf, Tom, & Carville, Olivia. (2018, June 20). Mark Zuckerberg is on the brink of becoming richer than Warren Buffet. Bloomberg, Retrieved from https://www.bloomberg.com/news/articles/2018-06-20/zuckerberg-on-brink-of-eclipsing-buffett-with-facebook-at-200

Meyer, David. (2018, April 6). Facebook decides now's not a great time to harvest patients' medical data. *Fortune*, Retrieved from http://fortune.com/2018/04/06/facebook-medical-data-sharing-hospitals

Meyer, Michelle (2014, June 29). How an IRB could have legitimately approved the Facebook experiment, and why that may be a good thing. *The Faculty Lounge*. Retrieved from http://www.thefacultylounge.org/2014/06/how-an-irb-could-have-legitimately-approved-the-facebook-experimentand-why-that-may-be-a-good-thing.html

Miller, Andrea. (2017). *Protocological violence and the colonial database. Antipode blog Intervention Symposium, 1-6. Symposium conducted at the annual meeting of the American Association of Geographers in San Francisco, CA*. Retrieved from https://radicalantipode.files.wordpress.com/2017/05/6-andrea-miller.pdf

Mirabal, Nancy. (2009). Geographies of displacement: Latina/os, oral history, and the politics of gentrification in San Francisco's Mission District. *The Public Historian, 31*(2), 7–31.

Pew Research Center. (2016, November). Social media update 2016. Retrieved from http://www.pewinternet.org/2016/11/11/social-media-update-2016/

Robinson, David, & Koepke, Logan. (2016). Stuck in a pattern: Early evidence on "predictive policing" and civil rights. *Team Up Turn* (Report). Retrieved from https://www.teamupturn.org/static/reports/2016/stuck-in-a-pattern/files/Upturn_-_Stuck_In_a_Pattern_v.1.01.pdf

Rosdail, Jennifer. (2014). The quad: A newly defined meta hood. Retrieved from http://www.jenniferrosdail.com/the-quad-is-born/

Rosenblatt, Joel. (2016, October 26). Is Facebook's facial-scanning technology invading your privacy rights? *Bloomberg*, Retrieved from https://www.bloomberg.com/news/articles/2016-10-26/is-facebook-s-facial-scanning-technology-invading-your-privacy-rights

Rose-Redwood, Reuben. (2008). From number to name: Symbolic capital, places of memory and the politics of street renaming in New York City. *Social & Cultural Geography, 9*(4), 431–452.

Rose-Redwood, Reuben, Alderman, Derek, & Azaryahu, Moaz. (2010). Geographies of toponymic inscription: New directions in critical place-name studies. *Progress in Human Geography, 33*(4), 453–470.

San Francisco Department of Public Health. (2007). Eastern neighborhoods community health impact assessment. Retrieved from https://www.sfdph.org/dph/files/opp/HIAofEasternSanFranciscoNeighborhoods.pdf

Science FARE. (2016, October 19). Social science: Include social equity in California Biohub. *Nature, 538*, 371.

Solnit, Rebecca. (2013). Diary. *London Review of Books*, *35*(3), 34–35. Retrieved from https://www.lrb.co.uk/v35/n03/rebecca-solnit/diary

Stoler, Ann Laura. (2008). Imperial Debris: Reflections on ruins and ruination. *Cultural Anthropology*, *23*(2), 191–219.

Swanson, Heather Anne, Bubandt, Nils, & Tsing, Anna. (2015). Less than one but more than many: Anthropocene as science fiction and scholarship-in-the-making. *Environment and Society: Advances in Research*, *6*(1), 149–166.

Tenants Together & Anti-Eviction Mapping Project. (2014). The speculator loophole: Ellis Act evictions in San Francisco. Retrieved from http://www.antievictionmappingproject.net/speculatorloophole.html

Thatcher, Jim, O'Sullivan, David, & Mahmoudi, Dillon. (2016). Data colonialism through accumulation by dispossession: New metaphors for daily data. *Environment and Planning D: Society and Space*, *34*(6), 990–1006.

Tischler, Victoria, Rademeyer, Alison, & Vostanis, Panos. (2007). Mothers experiencing homelessness: Mental health, support and social care needs. *Health & Social Care in the Community*, *15*(3), 246–253.

Vaughan, Sue. (2016, January 25) The problem with the Google bus program. *48 Hills*, Retrieved from http://48hills.org/2016/01/25/problem-google-bus-program/

Walker, Richard. (2018). *Pictures of a gone city: Tech and the dark side of prosperity in the San Francisco Bay Area*. Oakland: PM Press.

Ward, Jackie. (2018, May 15). Nurses protest to drop 'Zuckerberg' from SF General's name. *KPIX*, Retrieved from https://sanfrancisco.cbslocal.com/2018/05/15/nurses-protest-to-drop-zuckerberg-from-sf-generals-name

Weise, Elizabeth. (2017, January 27). Mark Zuckerberg drops suits to force sale of Hawaiian lands. *USA Today*, Retrieved from https://www.usatoday.com/story/tech/news/2017/01/27/zuckerberg-drops-kauai-land-suits-hawaii-quiet-title/97152332

Wong, Julia. (2017, March 28). Uber diversity report paints overwhelmingly white, male picture. *Guardian*, Retrieved from: https://www.theguardian.com/technology/2017/mar/28/uber-diversity-report-white-male-women-minorities

Wong, Queenie. (2016, October 11). ACLU: Facebook, Twitter and Instagram aided police surveillance of protesters. *Silicon Valley*. Retrieved from https://www.siliconvalley.com/2016/10/11/aclu-facebook-twitter-and-instagram-aided-police-surveillance-of-protesters

"Turn your brand into a destination": toponymic commodification and the branding of place in Dubai and Winnipeg

Reuben Rose-Redwood, Maral Sotoudehnia and Eliot Tretter

ABSTRACT

The selling of naming rights to corporate sponsors has led urban policymakers to increasingly view the identities of public places as rent-generating assets to fund urban infrastructure. Yet few scholars have critically analyzed this emerging global trend of toponymic commodification and the seeking of "naming rent." Through a combination of archival research, on-site field observations, and semi-structured interviews, this study examines how the practice of toponymic commodification is transforming the cultural landscapes of contemporary cities by considering two naming rights programs: Dubai's Metro Naming Rights Initiative and the Sponsor Winnipeg Program. In each case, we explore the implications of commodifying public place names as well as the conflicting perceptions of such sponsorship programs. In doing so, the present study illustrates how the selling of naming rights is reshaping the built environment into a space of symbolic/economic capital transformations as brands become destinations and public places are reconceived as marketing opportunities.

> Turn your brand into a destination … Welcome to the ultimate branding and marketing opportunity. With Dubai Metro Naming Rights, you can put your brand on a Dubai Metro station of your choice, or one of the two lines of the Dubai Metro Network (Roads and Transport Authority of Dubai as cited in Goldsbie, 2008).

> The City of Winnipeg offers a number of highly visible facilities and services for your company brand. From bridges and overpasses to parkades, web sites and parking meters, for the first time ever the City of Winnipeg is making available a number of high traffic and visible facilities and city properties for sponsorship (City of Winnipeg, (2015 [2009]).

Introduction

In the spring of 2008, the Roads and Transport Authority (RTA) of Dubai took out a full-page advertisement in both *Business Week* and *The Economist* promoting the Dubai

Metro Naming Rights Initiative, which sought to provide corporations with "the ultimate branding and marketing opportunity" by selling the naming rights for metro stations in order to generate revenue for Dubai's new mass transit system (Goldsbie, 2008). The ad pitched such opportunities to potential corporate sponsors as a form of "immersive marketing" that could literally turn a sponsor's "brand into a destination." At the same time, the City of Winnipeg, in Canada, was also in the midst of developing its own naming rights sponsorship program called "Sponsor Winnipeg" with the hope of tapping into an expansive new revenue stream without raising property taxes. Originally approved by the city council in 2007, Winnipeg's sponsorship program was officially launched in 2009 to promote the use of naming rights, and other related sponsorships, to fund public infrastructure. Although the selling of naming rights predates the Great Recession (the slowdown in the global economy that followed the financial crisis of 2007–2008), the economic downturn led an increasing number of municipalities to consider the sale of naming rights as part of a broader entrepreneurial strategy of neoliberal urban governance (Hopper, 2012).

Despite the proliferation of naming rights agreements in cities around the world, there has been surprisingly little geographical research on the commodification of urban toponymies. At present, much of the existing literature on naming rights has emerged from the fields of law, economics, marketing, and communications (Bartow, 2007; Blackshaw, 2012; Blocher, 2007; Boyd, 2000; Burton, 2008; Leeds, Leeds, & Pistolet, 2007). Urban scholars have examined city branding and place marketing as part of broader discussions of contemporary urbanism (Dinardi, 2017; Dinnie, 2011; Kavaratzis & Ashworth, 2005; for general discussion of the geographies of branding, see Pike, 2011, 2015). However, this focus on the "city-as-brand" has thus far not been accompanied by a detailed examination of the practices of "toponymic branding," or what recent studies have referred to as *toponymic commodification* – that is, the commodification of place names as a neoliberal strategy of urban place-making (Light & Young, 2015; Medway & Warnaby, 2014; also, see; Rose-Redwood, 2011; Rose-Redwood, Alderman, & Azaryahu, 2010; Vuolteenaho & Ainiala, 2009).

Through a combination of archival research, on-site field observations, and semi-structured interviews, this study examines how the practice of toponymic commodification is transforming the cultural landscapes of contemporary cities by considering two ongoing naming rights programs: Dubai's Metro Naming Rights Initiative and the Sponsor Winnipeg Program. In both cases, the process of toponymic commodification extends market relations to the production of place-identities by rendering the names of places into a form of economic capital (i.e. marketable assets). As we discuss in the next section, the selling of naming rights can be characterized as a rent-seeking practice that transforms the symbolic capital of place into the economic capital of the "toponym-as-commodity" (Rose-Redwood & Alderman, 2011; p. 3; also, see Light & Young, 2015). The present study examines the symbolic/economic capital transformations associated with the selling of naming rights from a critical geographic perspective. By describing the geographies of naming rights in two very different cities, we illustrate how important it is to situate political economic analyses of urban toponymies within the geographical contexts in which toponymic commodification takes place. Despite differences between the naming rights programs in Dubai and Winnipeg, putting the two cases into dialogue highlights how toponymic commodification is increasingly becoming commonplace globally and normalized as a "common sense" entrepreneurial strategy of neoliberal urban governance.

Neoliberal urbanism and the political economy of urban naming rights

The naming of places has long been associated with the appropriation of urban space by political, economic, and cultural elites as a means of reshaping the spatial identities of places in the image of wealth and power (Berg & Vuolteenaho, 2009; Creţan & Matthews, 2016; Giraut & Houssay-Holzschuch, 2016; Rose-Redwood et al., 2010). Buildings, streets, squares, neighborhoods, and entire cities have often been named in honor of political leaders, property owners, and other wealthy patrons to produce the taken-for-granted spaces of everyday urban life. Place naming plays an important role in the construction of *place-identities* and thus contributes to the making of cultural landscapes by inscribing culture in place. Consequently, the act of place renaming not only changes geographical imaginaries of place; it also reshapes the cultural landscape itself since place naming is a world-making practice (Rose-Redwood, Alderman, & Azaryahu, 2018).

The commodification of naming rights for public places is a relatively new urban policy trend, yet history is replete with "the corporate control of toponyms" (Post & Alderman, 2014, p. 84), from the naming of company towns during the nineteenth century to the gated communities of the contemporary era. Although legacy gifts date back centuries, philanthropic naming grew in popularity at the turn of the twentieth century (Burton, 2008). During the 1970s, however, a subtle shift occurred in the political economy of naming rights, which continues to have significant reverberations today. In 1972, the naming rights for the Buffalo Bills' new football stadium were leased to the Rich Products Corporation as part of a 25-year agreement whereby the company would pay $60,000 (USD) per year in exchange for securing the right to name the arena "Rich Stadium" (OMICS International, 2014). There is a long history of naming places after wealthy philanthropists as a symbolic gesture of gratitude for a significant gift or donation; however, the naming of Rich Stadium was one of the first instances in which the name of a professional sports arena, or any other place, was purchased as a *time-limited commodity.*

The selling of naming rights for professional sports arenas has now become a lucrative, multi-billion dollar global industry. According to a recent report, total spending on corporate sponsorships globally in 2017 was an estimated $62.7 billion (USD), over a third of which consisted of sponsorship spending in North America alone (ESP Properties, 2018). The majority of sponsorships are in the professional sports sector, and, since the late-1990s, colleges, universities, and even some high schools have also jumped onto the naming rights bandwagon to provide a revenue source for their sports facilities and programs (Blocher, 2007; Weinberg, 2003). A growing number of municipal governments have likewise embraced the selling of naming rights for public places (Hopper, 2012), and, in 2003, the Las Vegas Monorail Company sold the naming rights to a monorail station and train line for an estimated $50 million (USD) over 12 years (Hirsch, 2003). Subsequently, numerous transit authorities have followed Las Vegas's lead in commodifying the naming rights for metro stations (e.g. Chicago, Delhi, Dubai, London, Madrid, Mumbai, New York,

Philadelphia, Riyadh), although some metro naming rights proposals have encountered legal complications (Los Angeles) or lack of interest from potential sponsors (Boston). Nevertheless, municipal and transit-related naming rights programs gained momentum during the Great Recession as many cash-strapped cities were desperate for revenue at a time of fiscal austerity.

Since the 2007–2008 economic crash, many urban and economic geographers have examined the "neoliberal nostrums" of post-crisis austerity politics and policies (Peck, Theodore, & Brenner, 2013, p. 1092; also, see Rossi, 2017). The trenchant rolling-back of financial support and services by senior levels of government has left municipalities with little room to respond to mounting infrastructural deficits amidst growing urban populations and increased service demands. Municipal governments are thus encouraged to take financial matters into their own hands by securing direct and indirect revenues from sources other than taxes or fees. Through such entrepreneurial strategies of neoliberal urban governance, many municipalities have sought to marketize existing and proposed public assets in whatever way possible.

Leitner, Sheppard, Sziarto, and Maringanti (2007) argue that the "propagation of neoliberal discourses, policies, and subjectivities ... [has] given rise to [a new form of] neoliberal urbanism" (p. 4). Their argument builds upon an extensive body of urban scholarship that has connected the transition towards increased inter-urban competition and entrepreneurial urban governance to a series of spatially-grounded urban policies and initiatives since the 1970s (Brenner & Theodore, 2002; Harvey, 2005). Chief among these are city branding and marketing strategies as well as other efforts to explicitly tether urban cultural infrastructure to economic growth (Tretter, 2009). We contend that urban naming rights programs are another, largely underexplored, aspect of rolling out the policy agenda of neoliberal urbanism, which is currently remaking the cultural geographies of cities. We are not suggesting that there is a singular neoliberal urban agenda that has been uniformly adopted globally, but neoliberal urban policy ideas do travel through "global circuits of knowledge" (McCann, 2011), and the selling of naming rights to public places has now become part of the neoliberal "toolkit" of urban governance that a growing number of city governments are embracing. In order to better understand the process of producing toponymy-as-commodity, we must delve more deeply into the interrelations of naming, property, forms of capital, and rent-seeking.

The commodification of naming rights has the effect of privileging the potential economic exchange value of urban toponyms over other potential non-market values associated with urban place naming (Rose-Redwood et al., 2010). Through naming rights agreements, corporations and wealthy individual sponsors are granted the power to take ownership over, even if just in name, parts of the cityscape by branding them with their names or logos. Transforming public place names into potential rent-generating assets, city politicians and administrators are directly trading on a place's potential symbolic value, particularly as a site of public visibility, by transforming the symbolic capital of place naming – i.e. the prestige, distinction, and legitimacy associated with public commemoration and honoring – into a form of economic capital (i.e. a financial asset with an ongoing stream of revenue). Within this context, *symbolic capital* refers to the power that different social actors acquire through the perception of prestige or distinction (Bourdieu, 1990). Naming a place after a person (or corporate entity) is a form of public honoring that bestows social distinction, and the honoree can be said to accumulate symbolic capital through this act of public recognition (Alderman, 2008; Duminy, 2014; Post & Alderman, 2014; Rose-Redwood, 2008). *Economic capital*, embodied in a socially-recognized monetary form, is also accumulated and its uneven distribution has a substantial influence over how power is exercised by different social groups. Importantly, Bourdieu (1986) argues that symbolic capital can be converted into economic capital (and vice versa), and this transformational process is a significant aspect of how social hierarchies are reproduced. The selling of naming rights, we argue, is a novel illustration of the symbolic/economic capital transformation (Figure 1). In the case of toponymic commodification, the symbolic value of an urban toponym is monetized and treated as a commodity. When a sponsor buys – or, more accurately, leases – the right to name a place, the symbolic capital of urban place naming becomes part of an economic transaction.

Like all forms of property, the ability to trade in naming rights emerges, first and foremost, because of the capacity of someone to claim ownership of something and to ensure this control through some form of "legitimately" recognized source of social power. Property rights, however, always involve something more than just having and maintaining bare possession over something. Instead, the condition of possibility for having a system that allows property to be treated as assets – that is, as tradable goods – entails the ability to make something scarce by withholding it and restricting its use

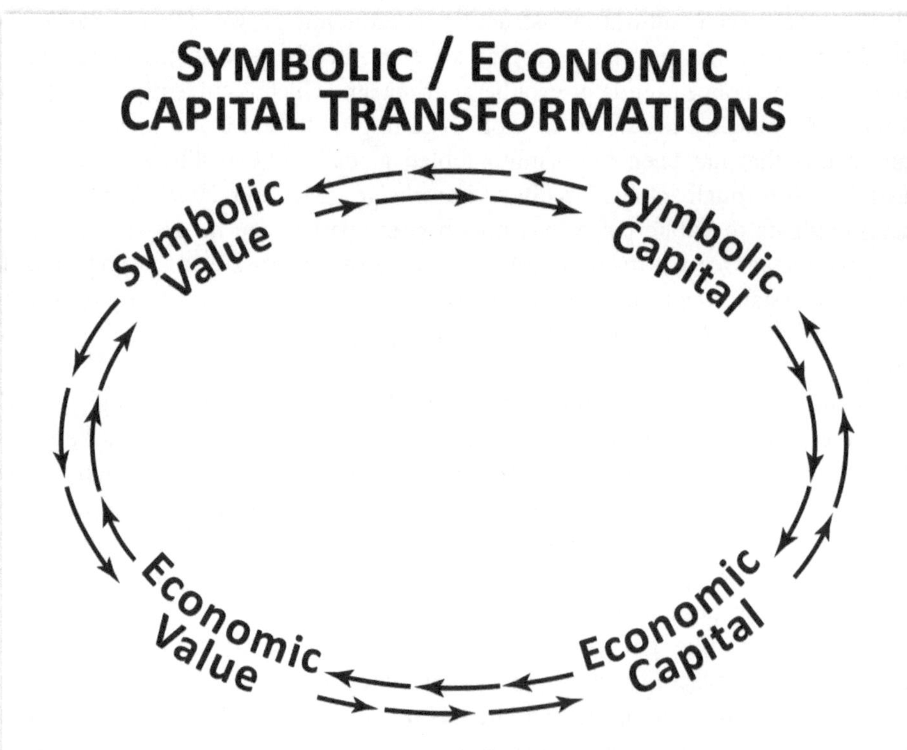

Figure 1. The transformations of symbolic and economic capital (artwork by Robin Poitras).

(May, 2013). The law plays a central role in rendering goods scarce and exchangeable, since it is through legal mechanisms that property rights are created and enforced. Imposing legal limitations on use grants certain parties the right to use something, transfer this right to someone else, or, often most importantly, transfer the profits that come from the use to another party in exchange for something else.

As in other markets, the transfer of naming rights for places includes buyers and sellers and is contingent upon the vagaries of supply and demand. If there are a limited number of locations with particular features (i.e. centrality, size, historical value, etc.), then these will likely command a higher price than other features. When there is more or less interest in securing a naming right, this will influence the price that a buyer pays and a seller is able to charge. The market in place names is also very imperfect. On the one hand, there are only a limited number of buyers, whether these are individuals, families, or firms, who will pay to be granted the right to have their unique name affixed to a specific location for a fixed period of time. On the other hand, there are only a handful of sellers – often only one, such as a local government or a large institution – and this gives the seller the ability to be a price maker, not a price taker. In most cases, there is, in fact, a natural monopoly held by the seller (i.e. a situation where an institution, because of its unique geographical command and control, has the sole authority over the supply of a resource under its territorial jurisdiction). In this respect, any profit that is derived from the trading of a naming right is based on a monopoly price, which is determined by the buyer's willingness to pay and the seller's capacity to charge (Howard & King, 1988). The owner (or leasee) of a naming right is thereby granted a monopoly power – that is, the sole right to have their name, and only their name, officially associated with a place.

In addition to being governed by the forces of supply and demand in the marketplace, the selling of naming rights to corporate sponsors is sustained because of perceptions of its potential indirect influence on the profitability of firms. While control over the naming of a location may not directly impact a firm's productivity (i.e. it will not make laborers work faster or increase the efficiency of a firm's operations), it may function as a unique source of potential revenue because of the way that firms earn an economic rent or a surplus-profit by having access to, or trading on, a resource that is not available to their competition (Smith, 2000; Tretter, 2015). There is "naming rent," therefore, because the owner of the named property (whether public or private) is appropriating a portion of the profits that firms speculate that they earn from being granted an exclusive (naming) right (Bryan, 1985). Yet the perceived economic benefit to corporate sponsors may be increasingly in doubt, since there is evidence suggesting that "naming rights do not have a lasting impact on the profitability of the firms that buy them" (Leeds et al., 2007, p. 581).

Irrespective of the economic benefit of naming rights to corporate sponsors, the pursuit by urban governments and other institutions for "naming rents" continues to grow and is becoming more professionalized and refined. In many cases, naming rights sales are the result of ad hoc *agreements* between local governments and individual sponsors. However, an increasing number of cities have adopted formal naming rights *policies*, which codify the rules for individual naming rights agreements. Some cities have even developed naming rights *programs* as a means of proactively seeking to commodify urban toponyms. The recent shift towards monetizing place-identities as city assets is an important new development that deserves critical scholarly analysis. Following a discussion of the methodology we employed for this study, the remainder of the present article explores this global trend of toponymic commodification in Dubai and Winnipeg.

Methodology

The majority of studies examining toponymic commodification draw attention to specific naming practices, events, and cities, yet, as Light and Young (2015) observe, there are currently very few in-depth case studies of toponymic commodification, let alone comparative case study research. By investigating two distinct naming rights programs, this article considers the ways in which various stakeholders (i.e. government, industry, and residents) interpret and respond to toponymic commodification. By offering a detailed analysis of Dubai's and Winnipeg's naming rights programs, we follow Light and Young's (2015) call to explore the diverse global trajectories of urban toponymic commodification initiatives but also try to bring the two cases into dialogue to illuminate both the specific and general aspects of this global urban policy trend.

To examine the emerging geographies of urban naming rights, we present the results of two case studies of government-led naming rights programs in the cities of Dubai and Winnipeg. We selected these two case studies for several reasons. First, both cities have played a leading role in moving beyond ad hoc naming rights agreements by developing proactive naming rights sponsorship programs. Second, each naming rights campaign received considerable media coverage when it was launched during the Great Recession, which led urban policy-makers in other cities to view them as model cases to potentially be emulated elsewhere. Third, by bringing these naming rights initiatives into dialogue in two cities located over 11,000 km apart, on two different continents, we can illustrate some of the ways in which this global policy trend has had variegated local outcomes in diverse locales.

Data collection for both case studies took place between 2011 and 2012 and involved a combination of archival materials, on-site field observations in each location, and 37 semi-structured interviews (15 in Dubai, and 22 in Winnipeg). Interview participants were selected using a purposeful sampling strategy to ensure that the viewpoints of government officials, business leaders, and local residents were included in each case study. Since we were outsiders in Dubai and Winnipeg, the sampling of interviewees involved a combination of cold-calling, relying on gatekeepers, and drawing upon personal and professional networks.

Branding Dubai: taking toponymic commodification for a ride on Dubai's metro

In 2008, Dubai's Roads and Transport Authority (RTA) unveiled what it described as a "world first revenue model for public transport infrastructure" (Roads and Transport Authority of Dubai, 2012a). The program called for using revenue generated through the selling of naming rights for metro stations to offset the construction and maintenance costs of the new Dubai metro rail system (Al Theeb, 2008). According to the RTA, the program enabled the Government of Dubai to create infrastructure and provide essential services to residents without increasing user fees or taxes (Roads and Transport Authority of Dubai, 2012a). Within months of its launch, the RTA had already secured over 20 sponsors, including developers, investment groups, business conglomerates, banks, a telecommunications company, an airline company, and even two other government authorities (Table 1).

The RTA's website for the Dubai Metro Naming Rights Initiative is tailored specifically to entice potential sponsors with corporate brand exposure through immersive marketing opportunities. The website provides information about a government-led program, but it does so with the intent to sell the place-identities of public landmarks as a commodity. As the RTA's website explains, the naming rights program would "generate potential business opportunities" and be "a win-win partnership between the RTA and leading brands" (Roads and Transport Authority of Dubai, 2012b). Through such efforts, the RTA's naming rights program sought to extend the neoliberal strategy of public-private partnerships to the realm of urban place naming.

Like many of its architectural wonders, Dubai's metro system was initially unveiled to significant fanfare. *Arabian Business*, for instance, reported that "[t]he trains are good and the stations are what you'd imagine a Dubai Government project to be – spacious, shiny and glamorous (yes, they even have chandeliers)" (Egbert, 2009). Such descriptions of Dubai's metro system illustrate the prestige, or symbolic capital, associated with the Emirate's rapid transit infrastructure. Respondents, in fact, often commented on the metro's distinctiveness, and how the acquisition of rapid rail transit is a significant marker of Dubai's rise as a "global city." As one resident of Dubai explained:

> It's become this transportation landmark, so it makes Dubai more competitive with other cities. Now, every city in the Middle East is looking to Dubai because they have a Metro. It's considered as something so, so, European. People want to think they live in Paris now that there is a Metro. It's modern, cosmopolitan, like the city itself. In that sense, it has really benefited the city (Interview, May 2012).

Table 1. Naming rights sponsors of the Dubai metro naming rights initiative.

Sponsor Name	Type of Industry
Burjuman	Business Group
Burj Khalifa/Dubai Mall (Emaar)	Developer
Business Bay (Dubai Holding)	Investment Group
Creek (Dubai Holding)	Investment Group
DAMAC	Developer
Danube	Building Materials
Deira City Centre	Business Conglomerate
Dubai Airport Freezone	Government Authority
Dubai Health Care City (Dubai Holding)	Investment Group
Emirates	Airline
Emirates Tower (Dubai Holding)	Investment Group
Etisalat	Telecommunications
First Gulf Bank	Bank
Gulf General Investment Company (GGICO)	Investment Group
Jaddaf (Dubai Holding)	Investment Group
Jumeirah Lake Towers (Dubai Multi-Commodities Center)	Government Authority
Mall of the Emirates	Business Conglomerate
Nakheel	Developer
Nakheel Harbour and Towers	Developer
Noor Islamic Bank	Bank
Palm Deira	Developer
Sharaf DG	Investment Group
TECOM (Dubai Holding)	Investment Group
UAE Exchange (Jebel Ali)	Money Transfer Firm

Despite obvious admiration for the metro system itself, some government officials viewed metro station names as lacking symbolic "meaning" altogether, thereby legitimating the commodification of urban toponymy, with one participant in this study noting that:

> the station is not something that can have so much meaning . . . to, say, the guy [who] is living beside [it] . . . the name for them is x or y. I mean . . . the name itself is not maybe something which has serious impact on people. It's not the project itself. Like having the station here, or having the line here, or something like that. Here we are talking about just the station. [It] is there already and they are given a name. So the name could be x or y or z, or so (Interview, 2012).

From this standpoint, the metro station name selected is viewed as inconsequential so long as the monetary return is significant and the metro is operational.

Other local residents concurred with this governmental viewpoint. As one resident put it, "people don't give a shit what you name a metro station. You think people care? They don't" (Interview, 2012). This sentiment was echoed in other interviews as well. For instance, consider the excerpt below from a mother of two living in Dubai. When queried about the metro naming rights initiative, she responded by asking:

> why are you studying the metro names? At least the metro is there now. Other cities don't have the same idea, but maybe they will. But, here, in Dubai, nobody pays attention to these things [i.e. metro station names]. Especially not with the metro because it's not as important to people. It doesn't affect them. I think you are asking the wrong question. You should look at all the big projects, like the Palms, instead. Everybody loves learning about the Palm (Interview, 2012).

To some, therefore, the commodification of public place names was irrelevant to their daily lives, and if the selling of naming rights is what was required for the government to provide mass transit, then it was unnecessary to question this practice, let alone conduct an academic study on the topic. Toponymic commodification, and the corporatization of public space, "just becomes normal, like every other advertisement everywhere else," as another interviewee noted (Interview, 2012).

While most participants found the issue of corporate naming rights unremarkable, many who acknowledged the financial benefits of the metro naming rights program were not entirely comfortable with the corporatization of public place names. As one metro user explained:

> the naming rights for the stations, I think they help the government make money and bring some good, high profile, companies to Dubai. And those companies, they boost Dubai in the world, you know? If a bank sponsors the station, then people, tourists, they might think, "wow, this place has money," or something . . . But, actually, I think they shouldn't let corporations name everything (Interview, 2012).

This comment encapsulates a sentiment shared by most informants, many of whom had few, if any, qualms about the naming rights initiative but expressed some unease with the idea of a fully corporatized cityscape. Other interviewees, however, were more adamant in their opposition to the corporatization of urban public spaces. As one young man travelling to work in Dubai insisted:

> something's gotta be sacred. You can't sell out every time . . . there's gotta be a balance. There are a few places where you wouldn't want to name it the HSBC Gold Market. Ugh, really? I guess it wouldn't matter if it was still known in the town as the Gold Market. But you don't want to name it the Monsanto station, or Coca-Cola, or whatever. I think it's

> good to name a couple of places after historical people, places, things. Just anything to connect yourself to this place. Otherwise, it's completely synthesized ... Even if they could somehow show the story of Dubai and how crazy it is and where it's going in the future instead of just having the glitzy parts. It just needs something to root itself, a little. Just anything. People want substance and it definitely lacks substance (Interview, 2012).

Some participants, such as the one quoted above, suggested that the use of historical or commemorative place names could provide a platform for the Government of Dubai to showcase significant aspects of the city's culture and history to Emirati citizens, non-Emirati residents, and tourists traversing the city by rail. Yet, by leasing out the naming rights to public infrastructure, the RTA has privileged paying homage to corporate power instead of commemorating people, places, and events of historical significance.

Other participants who expressed concerns over the naming rights program highlighted the risks inherent to a sponsor's brand. As a young woman who often travels to Dubai observed, "there are all kinds of risks. You don't know if the company's reputation will change, or how it will change. Look at BP, or Enron. You just don't know what that name will mean in one, five, or ten years" (Interview, 2012). If the brand image suffers, then a station bearing the corporation's name, and, by extension, the government approving the naming rights agreement, could also be affected to the extent that they are perceived as part of that brand's asset. Moreover, if the RTA enters into a naming rights agreement with a sponsor that becomes embroiled in a scandal, other potential partners might be reticent to partake in the same corporate naming rights program. For instance, when it was revealed that some companies receiving federal bailout money from the U.S. government during the Great Recession were using taxpayer funds to pay millions of dollars for naming rights, this led to a temporary cooling of the naming rights market for sports arenas in the United States (Sandomir, 2010).

If the visibility of a corporate brand may pose risks to a sponsor's corporate image in some cases, the invisibility and banality of commodified urban toponyms has also led some branding experts to question whether the corporate sponsorship of public infrastructure is money well spent for corporations. As the CEO of an international branding firm based in the region maintained:

> the train project, in terms of the brands that have been used, or been associated with the train stations, I think that has been a mistake. I think it's cost a lot of money [for the corporations], and I don't think the brands have actually gained the recognition that they would have liked to see themselves get ... what the brands do when they badge themselves onto a station is they try to change people's natural description of a particular place. So, where a station may be station 5, or station Sheikh Zayed, or [the] station outside the Dubai Mall, or whichever station it is, you know, if somebody then goes and calls that Al-Utta station, or another station name, and they want to try and get people to use the name of the badge, the badge brand rather than the primary destination brand, I think there is a risk that that won't happen. And I think that's what's happened. I don't think any of the station brands have really gained significant recognition. I don't think RTA has been particularly good at helping those brands gaining recognition, either. I think there's a lack of awareness and a lack of promotion of the brands that have actually bought the naming rights to the stations (Interview, 2012).

Indeed, to our surprise, fieldwork observations confirmed that Dubai's metro terminals did not appear overly commercialized. Sponsors' brand names were displayed throughout the metro system, but they were not explicitly constructed around a corporate message or ad campaign, and visual evidence of corporate sponsorships was very subtle. At first sight, it would be difficult for a tourist or non-resident to distinguish corporate-sponsored metro station names from their unsponsored counterparts. Some sponsored names, such as the Mall of the Emirates station, have plaques explaining the sponsorship agreement (Figure 2). However, other corporate-sponsored names, such as Etisalat

Figure 2. Mall of the Emirates sponsorship sign in the Dubai Metro (photograph by Maral Sotoudehnia).

Figure 3. The Burk Khalifa/Dubai mall station (photograph by Maral Sotoudehnia).

Station (a local telecom company), Nakheel Harbour & Tower (Nakheel being one of the largest developers in Dubai), and the Burj Khalifa/Dubai Mall station, were barely noticeable (Figure 3). It was only through a detailed inquiry into the origin of each station's name that we determined which station names arose from naming rights sponsorships. To the extent that this reflects how corporate branding is becoming normalized in Dubai's cityscape, the everydayness of toponymic commodification may undercut the marketing potential that would lead corporate sponsors to purchase the naming rights for public infrastructure in the first place.

Sponsoring Winnipeg: commodification, commemoration, and the "value of honours"

On 14 November 2007, the Canadian city of Winnipeg approved a program called Sponsor Winnipeg, which paved the way for the city to pursue naming rights sponsorship opportunities for public facilities and services. The sponsorship initiative was proposed by the city's mayor, Sam Katz, and the city council passed the measure by a vote of 12 to 4 (City of Winnipeg, 2007). Prior to becoming mayor, Katz was a leading player in Winnipeg's sports and entertainment industry, and, as one of the primary owners of the Winnipeg Goldeyes baseball team, he was personally involved in the construction of the city's new baseball stadium, CanWest Global Park, in 1999, which was later renamed CanWest Park in 2008 and then renamed yet again as Shaw Park in 2011 once Shaw Communications acquired the naming rights following the bankruptcy of CanWest (Kirbyson, 2011). As mayor, Katz sought to bring the same "entrepreneurial spirit" to the governing of the city as a whole by reducing government regulations, cutting business taxes, freezing property taxes, and promoting public-private partnerships. The roll-out of the Sponsor Winnipeg Program was therefore part of a broader neoliberal urban policy agenda, and Mayor Katz insisted that it would "help us fix our crumbling roads, provide city services and keep our taxes competitive" (*CBC*, 2007).

Although several Canadian cities had already adopted naming rights policies, Winnipeg's naming rights sponsorship program was by far the most ambitious. "From bridges and overpasses to parkades, web sites and parking meters," announced the Sponsor Winnipeg website, "for the first time ever the City of Winnipeg is making available a number of high traffic and visible facilities and city properties for sponsorship" (City of Winnipeg, 2015 [2009]). Prospective sponsors were offered a virtual menu list of sponsorships to choose from, including everything from city parks, bridges, and transit services to swimming pools, community centers, and even a police helicopter. No city asset, it seemed, was off limits except perhaps City Hall itself, which was viewed as "sacred ground" by one of the local politicians who supported the program that we interviewed (Interview, 2011).

Right from the outset, opponents of Winnipeg's sponsorship program, such as City Councillor Jenny Gerbasi, cautioned that "you have to be careful that you don't go so far as to turn your city into one big commercial" (*CBC*, 2007). As the city began moving forward with launching its sponsorship program, some local residents raised concerns as well. For instance, one local commentator wrote a letter to the editor in the *Winnipeg Free Press*, arguing that:

> This program will destroy the value of honours a thankful citizenry bestows in recognition of real service (Stephen Juba Park), real sacrifice (Andrew Mynarski VC Junior High School) or real achievement (Milt Stegal Drive). Now having your name on a *public asset* could mean nothing more than that you paid to put it there. No honour in that and what an insult to all those who really did earn their honours (MacMillan, 2009, p. 11, italics added).

From this perspective, the naming of public places should be reserved for historical commemoration and honorific recognition of "real" service, sacrifice, and achievement. Reducing public place names to commodities that can simply be bought and sold was thus viewed as cheapening the symbolic value of urban toponymic landscapes. Yet not everyone shared this view. One city councillor responded to such arguments in the following manner:

> I don't think that speaks to how the general public view it ... To tell you this, *I can't think of somebody who is left to honor*. I mean, I don't think we're getting the names forward anymore like we used to for people to honor ... [in the past], you might have had a bridge named after somebody who didn't contribute to that ... there was never the monetary side that went with it. So, what I began doing now is looking at what are our *assets* that we can actually say are name-worthy (Interview, 2011, italics added).

As the process of toponymic commodification gained momentum, the very conception of name-worthiness, and the symbolic capital associated with it, appeared to be shifting from the commemorative logic of honorific recognition towards an economic logic of maximizing the rent that can be extracted from prospective sponsors, so much so that some city politicians could no longer "think of somebody who is left to honor" (Interview, 2011).

However, despite the city's efforts, the Sponsor Winnipeg Program has consistently failed to achieve the revenue targets that its promoters had championed at its inception. Mayor Katz and his allies had projected that sponsorships would generate $1.5 million (CAD) annually, yet the program's annual revenues have come nowhere near this goal. During its first year, Sponsor Winnipeg did not generate any revenue at all and actually cost the city $36,000 in operational expenses. The most revenue that the program has made in a single year was $396,200 in 2011, yet even this amount has not been sustained over time, with total sponsorship revenue dropping to $138,900 for 2015 (Kives, 2016). The first sponsorship agreements were made in 2010; however, none of these involved the sale of naming rights. When our research team conducted fieldwork in 2011, the first three naming rights deals had just been approved. Based on our interviews, it became evident that, at the time, there was a sharp divide between the majority on the city council (who adopted the naming rights policy) and the administrative staff (who were given the task of designing and implementing the program). As one leading administrator explained:

> the program hasn't generated sufficient revenues to pay for 2 km of road ... so the continued tension that exists right now is the expectation across the courtyard [at city hall] that it can still generate vast sums of money, and my own belief is that it's crazy and should just be treated as such .... As a general rule, I think it's better to accomplish something in life and have something named after you for that accomplishment instead of just buying your recognition (Interview, 2011).

Sponsor Winnipeg nevertheless moved forward despite such concerns, yet the sponsorship program would continue to be criticized for its "lackluster performance in generating revenue" in the years that followed (Taylor, 2016b).

The first naming rights agreement that was made as part of Sponsor Winnipeg took effect in May 2011, when the Akman and Shindleman families acquired the naming rights for a new children's program room at the Henderson branch of the Winnipeg Public Library for $25,000 (Figure 4). Mayor Katz was a close "friend and business associate" of Sandy Shindleman, President of Shindico Realty Inc., the real estate company that owns the shopping mall in which the Henderson Library is located (Lett, 2016). In 2009, the mayor and his allies on the city council approved an amendment to the lease for the Henderson Library at the request of Shindico, despite criticism that it would have been more cost effective for the city to purchase its own property for the library instead of continuing to pay rent. The amendment extended the lease from 10 to 20 years in exchange for renovations and expansion of the rental space for the library (City of Winnipeg, 2009). After receiving a lucrative, long-term lease from the city (worth close to $9 million in total rent), Shindico hired Akman Construction Ltd. to renovate and

Figure 4. The Akman/Shindleman children's program room in the Henderson Public Library (photograph by Reuben Rose-Redwood).

expand the Henderson Library (Akman Construction Ltd, 2017; Kives, 2009). Following the completion of the renovations, one of the new rooms was christened the "Akman/Shindleman Children's Program Room" in perpetuity for a sponsorship of $25,000, thereby becoming "the first naming right agreement to come before Council since the launch of Sponsor Winnipeg" (City of Winnipeg, 2011). During his time as mayor, Katz developed a reputation as someone who "frequently and shamelessly mixed city and personal business," so it is fitting that the first Sponsor Winnipeg naming rights deal would be with one of the mayor's friends and business associates. It would later be uncovered that "Katz had purchased a million-dollar home in Phoenix from the sister of an executive at Shindico Realty ... [with] no mortgage registered against it, suggesting either some sort of preferential financing by the vendor or a cash transaction" (Lett, 2016).

The next two naming rights sponsorships were for an ice rink and baseball diamond at the Southdale Community Centre. In 2011, a real estate company, Qualico, purchased the naming rights in perpetuity for the "Sage Creek Leisure Ice Rink" for $300,000 to promote its new Sage Creek development project, while real estate agent Tod Niblock acquired the naming rights for baseball diamond #2 (now officially the "Tod Niblock Baseball Diamond," see Figure 5). Subsequently, several additional naming rights sponsorship agreements were approved with sponsors such as TD Bank, Bockstael Construction, Kinsmen Club of Winnipeg, Raiders Hockey Inc., Casera

Figure 5. The Tod Niblock Baseball Diamond at the Southdale Community Centre in Winnipeg (photograph by John Lehr).

Credit Union, Gauthier Automotive Group, and the University of Winnipeg (Table 2; City of Winnipeg, 2017a). However, the city has struggled to entice prospective sponsors to purchase naming rights for public places. When asked about Winnipeg's sponsorship program, a leading figure in the city's business community responded by observing that:

> companies see much more effective ways in today's world to market their organization than to put their name on a *city asset* ... you know there is cynicism in terms of the city who has held on property taxes for years but has nickled and dimed some other revenue aspects, when it comes to user fees and other things, and I think the risk is to the perception of the public that it's just another money grab (Interview, 2011).

It should not be assumed, therefore, that a city's business leaders will be universally supportive of the roll-out of neoliberal urban policies, especially if the business community was not adequately consulted in the policy-making process.

Ironically, although Sponsor Winnipeg was designed to foster public-private partnerships as part of a neoliberal urban policy agenda, the city's sponsorship program has been criticized for adding yet another "layer of red tape and bureaucracy" for sponsorship proposals (Taylor, 2016a). One prominent critic of the program, City Councillor Russ Wyatt, has also argued that the city is "prostituting" itself by selling naming rights of public goods for "next to nothing" in terms of financial compensation (Taylor, 2016a). As Wyatt put it, "It was [Katz's] baby, but it's gone nowhere really in a big way. Let's just scrap it" (as quoted in Taylor, 2016a). In 2017, a decade after the program was originally approved, the city's administrative staff presented a "Business

Table 2. Naming rights sponsors of the Sponsor Winnipeg program.

Sponsor Name	Type of Industry
Akman/Shindleman Family (associated with Akman Construction Ltd. and Shindico Realty Inc.)	Individuals (Construction/Developer)
Bockstael Construction Ltd.	Construction Company
Casera Credit Union	Credit Union
Gauthier Automotive Group	Transportation Company
Kinsmen Club of Winnipeg	Community Group
Raiders Hockey Inc.	Sports Company
Sage Creek Development Corporation (Qualico)	Developer
Tod Niblock	Individual (Real Estate Agent)
Toronto-Dominion (TD) Bank	Bank
University of Winnipeg	Educational Institution

Case" for improving Sponsor Winnipeg based upon "the anticipated best-case financial benefits to be achieved" (City of Winnipeg, 2017b). The report recommended establishing a centralized office within the city's aptly-named Customer Service and Communications Division to "oversee" the sponsorship program, with $123,987 provided annually for the salary of a full-time employee to manage the program, plus an additional $30,000 to "be allocated annually … for an external contractor to valuate new or changing corporate assets" (City of Winnipeg, 2017c, p. 9). This recommendation was made based upon the claim that "there is a direct link between the resources allocated to the Program and the sponsor revenues generated" (City of Winnipeg, 2017c, p. 3). Yet, if past experience is any guide, basing decisions for naming rights programs on "best-case" scenarios may lead to false expectations as it did when the public was assured that the Sponsor Winnipeg Program would generate $1.5 million each year. Indeed, if the program were to only bring in $138,900 in a year, as it did in 2015, then the city would actually lose over $15,000 in net annual revenue by maintaining its "entrepreneurial" efforts to commodify the symbolic identities of urban public places.

The variegated geographies of toponymic commodification

Naming rights programs have become commonplace strategies of neoliberal urbanism over the past decade, but, as the juxtaposition of Dubai's and Winnipeg's experiences with selling naming rights illustrates, the geographies of toponymic commodification are variegated and have produced widely divergent outcomes in different urban contexts (for a discussion of "variegated neoliberalization" more generally, see Brenner, Peck, & Theodore, 2010). Placing Dubai's Metro Naming Rights Initiative in dialogue with Sponsor Winnipeg highlights how neoliberal policies have been adopted by both authoritarian and democratic regimes, which, despite their contrasting political ideologies, have similarly viewed the spatial identities of public places as potential rent-generating assets. In both cases, we have shown how the selling of naming rights has resulted in the conversion of symbolic capital into economic capital to serve as a revenue source for the construction and maintenance of public infrastructure.

As public-private partnerships, naming rights sponsorship programs are often touted as an undifferentiated entrepreneurial solution to urban fiscal problems. However, not all naming rights initiatives have been equally embraced. For instance, the first 13 naming rights deals for Dubai's metro stations generated 1.8 billion Dirhams (equivalent to approximately $489 million USD) (Al Theeb, 2008), whereas, between 2009 and 2015, Sponsor Winnipeg only brought in $1.87 million (CAD) ($1.35 million USD) in total sponsorship revenues (Pursaga, 2016). While Dubai's naming rights initiative is reported to have offset 60 percent of the new metro's operation and maintenance costs (Sophia, 2014), Winnipeg's sponsorship program has struggled to secure sponsors and has consistently failed to live up to expectations (Kives, 2016). These contrasting outcomes complicate a straightforward account of the effect that the selling of naming rights is having on the political economy of urban development, which underscores the importance of examining the specificities of toponymic commodification in different geographical locales.

In the cases of Dubai and Winnipeg, there are at least two key differences between their respective naming rights programs that likely influenced how each initiative developed: scope and visibility. First, the scope of Dubai's naming rights program was limited to metro stations and rail lines in the city's newly constructed, hyper-modern mass transit system. From the very outset, corporate sponsors lined up to acquire naming rights for metro stations, the names of which would appear not only on signage throughout the metro system itself but also on highway signs, maps, digital locational apps, tourist materials, etc. To the extent that metro stations become incorporated into the wayfinding practices of urban inhabitants, the naming of these new places provided an opportunity for corporations to literally inscribe their brands within the geographical imaginary of the city – that is, to turn their brands into destinations as locational nodes within the urban toponymic landscape. By contrast, Winnipeg's sponsorship program includes nearly every public asset that the city owns – from parking meters and library reading rooms to recreation centers and bridge overpasses. Despite this extensive inventory of potential naming rights opportunities, Winnipeg has only acquired a few sponsorships over the past decade, all of which have been for relatively modest renovations or maintenance of existing infrastructure. The scope of Sponsor Winnipeg's naming rights inventory may be far more extensive than that of Dubai, but the latter's targeted focus on major nodal points within an urban transportation network proved far more effective in generating opportunities to transform the symbolic capital of urban toponymy into economic capital for urban infrastructural development. Yet other metro naming rights initiatives have floundered in cities such as Boston (Grillo, 2014), so there must clearly be other factors at play in shaping the economic viability of naming rights programs as well.

Another major difference between the two cases of naming rights in Dubai and Winnipeg involves the issue of global visibility. With a population of over 3 million inhabitants, Dubai has sought to brand itself as a global city by dazzling the world with the construction of over-the-top, record-breaking, sensational landmarks, such as the Burj Dubai/Khalifa (the tallest building in the world) and Dubai Mall (the world's largest shopping mall by total area), among others. Dubai's metro system is also the "Arabian Peninsula's first urban train network" (BBC, 2009), and it has already inspired Saudi Arabia to launch a similar metro naming rights program in Riyadh (Graves,

2017). By comparison, Winnipeg is a relatively small Canadian city with a population of less than a million that does not have the same level of visibility for global capital or the tourism sector. Consequently, there is less demand for naming rights from potential sponsors and some local business leaders have even questioned the economic rationale behind Winnipeg's naming rights program. One of the main reasons why many professional sports arenas can garner such high prices for naming rights is because each time there is a sporting event the sponsor's name is potentially seen by millions of sports fans. Visibility is therefore a key factor in shaping the market for naming rights, and smaller cities such as Winnipeg are thus at a significant disadvantage to larger and more prominent cities like Dubai given the more limited visibility of the former's public assets in comparison to the latter.

In both Dubai and Winnipeg, our research team encountered interviewees who supported the selling of naming rights for public places while others expressed concern over whether toponymic commodification would result in the over-commercialization of public space. Many, however, were indifferent to the issue altogether. Such indifference often arises due to the banality of "prosaic names that barely attract any attention whatsoever" (Berg, 2011, p. 17). To the extent that commodified toponymies become normalized in everyday urban life, the "banal namescapes" associated with municipal naming rights programs will likely consist of "names that raise little or no political response" (Berg, 2011, p. 17). However, the apparent normality of a place name can change dramatically if a sponsor's public image as a responsible corporate citizen or private philanthropist falls into disrepute. Moreover, sponsors also run a risk by associating their brand with particular public places, since they cannot fully control the potential negative connotations related to events that may occur at those locations.

Although it is important to consider the specificities of each naming rights program as a unique case of toponymic commodification, urban policies that are developed in one city may come to be viewed as a model for other cities to emulate. As noted above, Dubai's Metro Naming Rights Initiative has already become a model for financing Riyadh's metro system. Similarly, when Sponsor Winnipeg was initially launched, it received considerable media coverage in the Canadian press, leading urban policymakers across the country to wonder whether they should follow Winnipeg's lead in pursuing the sale of naming rights for their municipal assets. As these cases illustrate, the development of naming rights policies and programs does not occur in a vacuum, so we must be attentive to both the specificity *and* relationality of urban policymaking in diverse geographical contexts (McCann, 2011).

Conclusion

The commodification of naming rights for urban public places has become part of the entrepreneurial toolkit of neoliberal urban governance, enticing urban policymakers with the prospect of generating non-tax revenue to finance urban infrastructure and services. It has also raised concerns among critics that naming rights policies will result in the over-commercialization of public space. In practice, our research shows that the revenue generated from urban naming rights programs does not always meet the optimistic expectations of policymakers, nor do such programs result in excessively commercialized public spaces in all cases, either. This study therefore underscores how the contextual specificities of place matter when considering the practices of toponymic commodification, since the outcomes of naming rights programs vary considerably across different cities.

In the present study, we have extended Bourdieu's (1986) theory of symbolic/economic capital transformations by illustrating how the selling of naming rights can be understood as an exchange of symbolic capital for economic capital (from the seller's perspective) and economic capital for symbolic capital (from the buyer's standpoint). In the process, the seller may also hope to enhance their own symbolic capital by associating themselves with a prominent sponsor, while the buyer of a naming right may view the transaction as part of their marketing efforts with the aim of generating future economic returns. In the case of naming rights, then, the distinction between "symbolic" and "economic" capital inherited from Bourdieu becomes blurred, which requires a re-evaluation of Bourdieu's notion of symbolic capital.

For Bourdieu, the symbolic capital that accompanies the power of naming is largely based upon the capacity of a ruling faction to maintain their social position outside of market valuation. He argues, for instance, that "[i]n the symbolic struggle for … the monopoly over legitimate naming …. symbolic order is not formed in the manner of a market price" (Bourdieu, 1989, p. 21). Yet we have shown in this study how the process of monetizing the symbolic capital associated with urban place naming reconfigures the symbolic value of public recognition into a more explicitly economic affair. In other words, by emplacing Bourdieu's framework within current urban cultural geographies, we reveal a shift in which the symbolic order *is* being formed precisely in the "manner of a market price." Of course, prior forms of honorific and commemorative place naming generally maintained the symbolic value of social hierarchies and should therefore not be romanticized through a politics of nostalgia as being bestowed only upon the worthiest. Yet the recent development, and widespread diffusion, of contractual naming rights agreements for public spaces goes a step beyond even the gift economies of philanthropic naming by transforming public place names into financial assets. In this case, the monopoly over legitimate naming still forms the basis for public recognition, but it also establishes the condition of possibility for the appropriation of an economic rent, making toponymic commodification a new frontier in value extraction.

The commodification of urban toponymies is one facet of the privatization of public space that has accompanied the neoliberal turn towards public-private partnerships as the foundation of urban governance generally and the provision of urban infrastructure in particular. By framing the issue of naming rights solely in economic terms, other potential non-economic values of public toponymies are devalued in the process. In particular, the opportunity to use public place naming as an arena in which to challenge existing social and cultural hierarchies is lost. The naming of public places has the potential to create a more socially-inclusive cultural landscape that more equitably recognizes socially and culturally marginalized groups. This can take the form of naming places in honor of racialized minorities who have been excluded from the spaces of public commemoration, creating a more equitable gender distribution in urban namescapes, or reclaiming indigenous toponymies that have been placed under erasure by settler-colonial societies. While we should be cautious about how emancipatory these more inclusive forms of commemorative place naming actually are – since they too can be enrolled in neoliberal place-marketing efforts – they can nevertheless serve as useful reminders that there are indeed alternatives to the commodified toponymies of neoliberal urbanism.

Disclosure statement

No potential conflict of interest was reported by the authors.

Funding

This work was supported by the Social Sciences and Humanities Research Council of Canada [410-2011-365].

References

Akman Construction Ltd. (2017). Henderson library. Retrieved on June 7, 2017, from http://www.akman.ca/our-projects/henderson-library

Al Theeb, Alia. (2008, December 22). Dubai raises Dh1.8bn from metro-naming rights and funding. *Gulf News*. Retrieved on June 6, 2017, from http://gulfnews.com/news/gulf/uae/traffic-transport/dubai-raises-dh1-8b-from-metro-naming-rights-and-funding-1.71114

Alderman, Derek. (2008). Place, naming and the interpretation of cultural landscapes. In Brian Graham & Peter Howard (Eds.), *The Ashgate research companion to heritage and identity* (pp. 195–213). Aldershot: Ashgate.

Bartow, Ann. (2007). Trademarks of privilege: Naming rights and the physical public domain. *UC Davis Law Review*, *40*, 919–970.

BBC. (2009, September 11). Will metro change Dubai care culture? *BBC News*. Retrieved on May 26, 2018, from http://news.bbc.co.uk/2/hi/middle_east/8247330.stm

Berg, Lawrence. (2011). Banal naming, neoliberalism, and landscapes of dispossession. *ACME: an International E-Journal for Critical Geographies*, *10*(1), 13–22.

Berg, Lawrence, & Vuolteenaho, Jani (Eds.). (2009). *Critical toponymies: The contested politics of place naming*. Burlington, VT: Ashgate.

Blackshaw, Ian. (2012). *Sports marketing agreements: Legal, fiscal and practical aspects*. The Hague, Netherlands: Springer.

Blocher, Joseph. (2007). School naming rights and the first amendment's perfect storm. *Georgetown Law Journal*, *96*(1), 101–159.

Bourdieu, Pierre. (1986). The forms of capital. In John Richardson (Ed.), *Handbook of theory and research for the sociology of education* (pp. 241–258). New York: Greenwood Press.

Bourdieu, Pierre. (1989). Social space and symbolic power. *Sociological Theory*, *7*(1), 14–25.

Bourdieu, Pierre. (1990). *In other words: Essays towards a reflexive sociology*. Stanford: Stanford University Press.

Boyd, Josh. (2000). Selling home: Corporate stadium names and the destruction of commemoration. *Journal of Applied Communication Research*, *28*(4), 37–41.

Brenner, Neil, Peck, Jamie, & Theodore, Nik. (2010). Variegated neoliberalization: Geographies, modalities, pathways. *Global Networks*, *10*, 182–222.

Brenner, Neil, & Theodore, Nik. (2002). Cities and the geographies of "actually existing neoliberalism." In Neil Brenner & Nik Theodore (Eds.), *Spaces of neoliberalism: Urban restructuring in North America and Western Europe* (pp. 349–380). Malden, MA: Blackwell Publishers.

Bryan, Richard. (1985). Monopoly in Marxist method. *Capital & Class*, *9*(2), 72–92.

Burton, Terry. (2008). *Naming rights: Legacy gifts and corporate money*. Hoboken, NJ: Wiley.

CBC. (2007, November 9). Winnipeg to sell naming rights to civic facilities, services. *CBC News*. Retrieved on June 9, 2017, from http://www.cbc.ca/news/canada/manitoba/winnipeg-to-sell-naming-rights-to-civic-facilties-services-1.645944

City of Winnipeg. (2007, November 21). The City of Winnipeg sponsorship policy file FR-2. Council minutes. Retrieved on June 6, 2017, from http://winnipeg.ca/sponsorwinnipeg/pdfs/SponsorshipPolicy.pdf

City of Winnipeg. (2009, November 25). Lease amendment - Henderson Library - unit 1, 1050 Henderson Highway (North Kildonan Ward) eFile FI-2.3(33). Council minutes. Retrieved on June 9, 2017, from http://clkapps.winnipeg.ca/DMIS/ViewDoc.asp?DocId=9737&SectionId=240688&InitUrl=/DMIS/Documents/c/2009/m9737

City of Winnipeg. (2011, May 18). Sponsor Winnipeg–sale of naming rights–children's program room, Henderson Library (North Kildonan Ward). Agenda Policy Committee. Retrieved on June 9, 2017, from http://clkapps.winnipeg.ca/DMIS/ViewDoc.asp?DocId=11027&SectionId=277201&InitUrl=/DMIS/Documents/epc/2011/a11027

City of Winnipeg. (2017a, March 6). Appendix V: Sponsorships. Standing Policy Committee on Protection, Community Services and Parks regular meeting agenda. Retrieved on June 9, 2017, from http://clkapps.winnipeg.ca/DMIS/ViewDoc.asp?DocId=15969&SectionId=460340&InitUrl=/DMIS/Documents/pcs/2017/a15969

City of Winnipeg. (2017b, March 6). Sponsor Winnipeg business case. Minutes–Standing Policy Committee on Protection, Community Services and Parks. Retrieved on June 9, 2017, from http://clkapps.winnipeg.ca/DMIS/ViewDoc.asp?DocId=15983&SectionId=460661&InitUrl=/DMIS/Documents/pcs/2017/m15983

City of Winnipeg. (2017c, March 6). Sponsor Winnipeg business case, attachment 1. Standing Policy Committee on Protection, Community Services and Parks regular meeting agenda. Retrieved on June 9, 2017, from http://clkapps.winnipeg.ca/DMIS/ViewDoc.asp?DocId=15969&SectionId=460335&InitUrl=/DMIS/Documents/pcs/2017/a15969

City of Winnipeg. (2015 [2009]). Sponsor Winnipeg opportunities. Sponsor Winnipeg Program. Retrieved from http://winnipeg.ca/sponsorwinnipeg/opportunities.stm

Creţan, Remus, & Matthews, Philip. (2016). Popular responses to city-text changes: Street naming and the politics of practicality in a post-socialist martyr city. *Area*, *48*(1), 92–102.

Dinardi, Cecilia. (2017). Cities for sale: Contesting city branding and cultural policies in Buenos Aires. *Urban Studies*, *54*(1), 85–101.

Dinnie, Keith. (2011). *City branding: Theory and cases*. London: Palgrave MacMillan.from

Duminy, James. (2014). Street renaming, symbolic capital, and resistance in Durban, South Africa. *Environment and Planning D*, *32*(2), 310–328.

Egbert, Conrad. (2009, September 19). Dubai has finally built its own metro—Yet another first for this extraordinary city. *Arabian Business*. Retrieved on March 1, 2012, from http://m.arabianbusiness.com/track-time-12680.html

ESP Properties. (2018). What sponsors want & where dollars will go in 2018. New York: ESP Properties. Retrieved on May 23, 2018, from http://strivesponsorship.com/wp-content/uploads/2018/03/What-Sponsors-Want-Where-the-Sponsorship-Dollars-Will-Go-In-2018.pdf

Giraut, Frédéric, & Houssay-Holzschuch, Myriam. (2016). Place naming as dispositif: Toward a theoretical framework. *Geopolitics*, *21*(1), 1–21.

Goldsbie, Jonathan. (2008, May 1). Schooled by Dubai do. *Torontoist*. Retrieved on June 6, 2017, from http://torontoist.com/2008/05/schooled_by_dub

Graves, LeAnne. (2017, October 15). Riyadh metro to auction station naming rights. *The National*. Retrieved on May 23, 2018, from https://www.thenational.ae/business/economy/riyadh-metro-to-auction-station-naming-rights-1.667481

Grillo, Thomas. (2014, February 27). MBTA's effort to make millions on naming rights is derailed. *Boston Business Journal*. Retrieved on May 27, 2018, from https://www.bizjournals.com/boston/real_estate/2014/02/station-naming-rights-scheme-derailed.html

Harvey, David. (2005). *A brief history of neoliberalism*. New York: Oxford University Press.

Hirsch, Jerry. (2003). Nextel, Las Vegas link brands in monorail deal. *Los Angeles Times*, November 17. Retrieved on May 23, 2018, from http://articles.latimes.com/2003/nov/17/business/fi-vegas17

Hopper, Tristin. (2012, May 11). Cities look to naming rights to fill their coffers but success is uncertain. *The National Post.* Retrieved on June 15, 2017, from http://news.nationalpost.com/news/canada/cities-look-to-naming-rights-to-fill-their-coffers-but-success-is-uncertain

Howard, Michael Charles, & King, John Edward. (1988). *The political economy of Marx.* New York: NYU Press.from

Kavaratzis, Mihalis, & Ashworth, G.J. (2005). City branding: An effective assertion of identity or a transitory marketing trick? *Tijdschrift Voor Economische En Sociale Geografie, 96*(5), 506–514.

Kirbyson, Geoff. (2011, November 1). Shaw renames skyscraper, baseball park. *Winnipeg Free Press.* Retrieved on June 9, 2017, from http://www.winnipegfreepress.com/local/canwest-fades-from-city-skyline-113260474.html

Kives, Bartley. (2009, November 5). New library, no more rent: Councillors. *Winnipeg Free Press.* Retrieved on June 7, 2017, from http://www.winnipegfreepress.com/local/new-library-no-more-rent-councillors-69267742.html

Kives, Bartley. (2016, November 29). And now, a word from Winnipeg's (theoreical) corporate sponsors. *CBC News.* Retrieved on June 9, 2017, from http://www.cbc.ca/news/canada/manitoba/winnipeg-sponsorship-revenue-1.3872173

Leeds, Eva, Leeds, Michael, & Pistolet, Irina. (2007). A stadium by any other name: The value of naming rights. *Journal of Sports Economics, 8*(6), 581–595.

Leitner, Helga, Sheppard, Eric, Sziarto, Kristin, & Maringanti, Anant. (2007). Contesting urban futures: Decentering neoliberalism. In Helga Leitner, Jamie Peck, & Eric Sheppard (Eds.), *Contesting neoliberalism: Urban frontiers* (pp. 1–26). London: Guilford Press.

Lett, Dan. (2016, March 1). Katz remains true to form. *Winnipeg Free Press*, Retrieved June 7, 2017, from http://www.winnipegfreepress.com/local/katz-remains-true-to-form-370649251.html

Light, Duncan, & Young, Craig. (2015). Toponymy as commodity: Exploring the economic dimensions of urban place names. *International Journal of Urban and Regional Research, 39* (3), 435–450.

MacMillan, Rob. (2009, December 8). Letter to the editor. *Winnipeg Free Press*, p. 11. Retrieved on June 9, 2017, from https://newspaperarchive.com/winnipeg-free-press-dec-08-2009-p-11

May, Christopher. (2013). *The global political economy of intellectual property rights: The new enclosures?* London: Routledge.

McCann, Eugene. (2011). Urban policy mobilities and global circuits of knowledge. Toward a research agenda. *Annals of the Association of American Geographers, 101*(1), 107–130.

Medway, Dominic, & Warnaby, Gary. (2014). What in a name? Place branding and toponymic commodification. *Environment and Planning A, 46*(1), 153–167.

OMICS International. (2014). Rich products. Retrieved June 6, 2017, from http://research.omicsgroup.org/index.php/Rich_Products

Peck, Jamie, Theodore, Nik, & Brenner, Neil. (2013). Neoliberal urbanism redux? *International Journal of Urban and Regional Research, 37*(3), 1091–1099.

Pike, Andy (Ed.). (2011). *Brands and branding geographies.* Cheltenham: Edward Elgar.

Pike, Andy. (2015). *Origination: The geographies of brands and branding.* West Sussex, UK: Wiley-Blackwell.

Post, Chris, & Alderman, Derek. (2014). "Wiping New Berlin off the map": Political economy and the de-Germanisation of the toponymic landscape in First World War USA. *Area, 46*(1), 83–91.

Pursaga, Joyanne. (2016, November 28). More staff needed for Sponsor Winnipeg. *Winnipeg Sun.* Retrieved on May 25, 2018, from http://winnipegsun.com/2016/11/28/more-staff-needed-for-sponsor-winnipeg/wcm/4e086f41-0e93-4612-9a56-42f2ad710d48

Roads and Transport Authority of Dubai. (2012a). *Dubai metro naming rights initiative.* Retrieved February 12, 2012, from http://www.rta.ae/wpsv5/links/NamingRights/index.html

Roads and Transport Authority of Dubai. (2012b). Objectives. Retrieved on March 1, 2012, from http://www.rta.ae/wpsv5/links/NamingRights/objectives.html

Rose-Redwood, Reuben. (2008). From number to name: Symbolic capital, places of memory, and the politics of street renaming in New York City. *Social & Cultural Geography*, *9*(4), 432–452.

Rose-Redwood, Reuben. (2011). Rethinking the agenda of political toponymy. *ACME: an International Journal for Critical Geographies*, *10*(1), 34–41.

Rose-Redwood, Reuben, & Alderman, Derek. (2011). Critical interventions in political toponymy. *ACME: an International Journal for Critical Geographies*, *10*(1), 1–6.

Rose-Redwood, Reuben, Alderman, Derek, & Azaryahu, Maoz. (2010). Geographies of toponymic inscription: New directions in critical place-name studies. *Progress in Human Geography*, *34*(4), 453–470.

Rose-Redwood, Reuben, Alderman, Derek, & Azaryahu, Maoz. (2018). The urban streetscape as political cosmos. In Reuben Rose-Redwood, Derek Alderman, & Maoz Azaryahu (Eds.), *The political life of urban streetscapes: Naming, politics, and place* (pp. 1–24). New York: Routledge.

Rossi, Ugo. (2017). *Cities in global capitalism*. Malden: Polity.

Sandomir, Richard. (2010, January 6). What's in a naming right? Certainly not cash. *New York Times*. Section B, p. 11.

Smith, Tony. (2000). The systematic place of technological rents in Capital III. In Andrew Light & Mechthild Nagel (Eds.), *Race, class, and community identity* (pp. 117–132). New York: Humanity Books.

Sophia, Mary. (2014, September 18). Property developer Damac buys naming rights for Dubai Metro Station. *Gulf Business*. Retrieved from http://gulfbusiness.com/property-developer-damac-buys-naming-rights-dubai-metro-station

Taylor, Stephanie. (2016a, April 11). Winnipeg's corporate sponsorship program slammed. *Metro News*, Retrieved on June 7, 2017, from http://www.metronews.ca/news/winnipeg/2016/04/11/winnipeg-committee-talks-corporate-sponsorship.html

Taylor, Stephanie. (2016b, April 13). Winnipeg's corporate sponsorship program headed for review. *Metro News*. Retrieved on June 9, 2017, from http://www.metronews.ca/news/winnipeg/2016/04/13/winnipeg_s-corporate-sponsorship-program-to-be-reviewed-.html

Tretter, Eliot. (2009). The cultures of capitalism: Glasgow and the monopoly of culture. *Antipode*, *41*(1), 111–132.

Tretter, Eliot. (2015). Live music, intercity competition, and reputational rents: Austin, Texas the "live music capital of the world." *Human Geography*, *8*(3), 49–65.

Vuolteenaho, Jani, & Ainiala, Terhi. (2009). Planning and revamping urban toponymy: Ideological alterations in the linguistic landscaping of Vuosaari suburb, eastern Helsinki. In Lawrence Berg & Jani Vuolteenaho (Eds.), *Critical toponymies: The contested politics of place naming* (pp. 227–252). Burlington, VT: Ashgate.

Weinberg, Ari. (2003, March 24). Biggest college sports arena naming deals. *Forbes*. Retrieved on June 15, 2017, from https://www.forbes.com/2003/03/24/cx_aw_0320ncaa.html

City renaming as brand promotion: exploring neoliberal projects and community resistance in New Zealand

Robin A. Kearns and Nicolas Lewis

ABSTRACT

Proposals to change the names of entire urban centres are rare. We examine the case of Blenheim, New Zealand, where in 2016, representatives of local businesses campaigned for its renaming as Marlborough City, in recognition of the region's wine industry. Although defeated the proposal threatened to over-write established settlement history. It presumed to rename Blenheim under the aegis of New Zealand Inc., a shorthand for the pervasive yet nebulous economic nationalism that seeks to yoke all local and national identity to enhancing export growth. Drawing on media reports, we interpret this example of toponymic commodification as a neoliberalized project of place-making. Ironically, Blenheim and Marlborough are colonial names that displaced a long-established Māori name. The proposal highlights both the perversities and the deeply contested claims-making that often underlie and animate toponymic politics. Ultimately, it illustrates some of the limits of rights claimed under neoliberalism.

"The Name does not make the place.....the place makes the name"
"Blenheim is a perfectly good name and the wine industry does not need free advertising"
(Correspondents to *The Marlborough Express*, May 2016).

Proposals to change the names of settlements and topographic features are relatively common in New Zealand as it confronts the damage done to Māori senses of place by colonialism. Renaming entire towns and cities as tacit branding exercises are, however, as unusual in New Zealand as elsewhere. In this paper, we examine a campaign to rename the city of Blenheim "Marlborough City."[1] Launched in May 2016 by representatives of the local business community, the campaign sought to align the city's name not with its established Māori name, but rather with the colonial name of its surrounding region. This regional name – Marlborough – has built an enviable international reputation for its Sauvignon Blanc wines (Moran, 2016). The proposal sought to capitalise on a name that has appeared on wine bottles and associated with award-winning wineries for the last thirty years. Proponents claimed the change would yield extensive economic gains from tourism and other product branding effects.

Drawing on media reports, we examine the politics underlying this example of toponymic commodification (Medway & Warnaby, 2014). We interpret the proposed renaming as reflecting a neoliberalized vision of place-making within the context of a relentlessly economic form of nationalism. Three arguments are advanced: first, the initiative was class-based, deriving from the claims-making of the newly affluent beneficiaries of neoliberalism; second, it was made thinkable by the dominance of economic identity over socio-historical identity and the remaking of the self within contemporary New Zealand; and, thirdly, it built on the neoliberalizing logics of place branding. In short, we argue that certain groups sought to assert the primacy of market image and reason over heritage and place-attachment. This assertion amounted to launching a symbolic corporate takeover of the region's urban centre. That the proposal failed indicates an over-reach by the proponents and points to at least one set of limits to neoliberalism in what remains a politically contested present. In this sense, ours is an account of a local political contest over the rights to name which has ultimately led to a reassertion of the resilience of place and heritage.

In what follows, we review aspects of the politics of place branding and, in particular, the commodification of place and tradition associated with economic nationalism under a dominant neoliberalism in New Zealand. We then establish the contemporary context of changing place names in New Zealand before recounting the case for the name change and the resistance it met. We conclude that this short-lived attempt to alter the toponymic landscape amounts to an aspirational neoliberal act of exclusion and erasure of social histories by a new economic occupation of the region.

Branding place: naming, claiming and commodifying toponymy

Urban settlements have been named and, in so doing claimed, for as long as they have been established. Place names have provided a symbol of identity and a politics of affiliation through which to make sense of, and from, place. Under advanced capitalism, they offer a "brand" with which individual enterprises and collective, territorialised economic projects can be symbolically associated to create, renew or safeguard value (Evans, 2003). This link between existing place names and branding has been particularly potent with respect to tourism, food and drink industries which have linked place and product through place-branding, "co-marketing," and conceptions of "*terroir*" (Kavaratzis & Ashworth, 2008; see also Pike, 2011b).

Creating a new name to market a city and its businesses is a larger undertaking. Settlements have been renamed across the world to dissociate them from negative associations with historical events or their geographical neighbours; East Detroit was for example renamed Eastpointe in 1992 to dissociate the city from Detroit. Similarly, many Australian, South African, Polish and Ukrainian settlements have been renamed to take back historical names and identities after political occupations, or to erase the celebration of egregious acts of colonialism. Examples of renaming to take advantage of a positive brand, however, are rare. One example is that of Toyota City (Anttiroiko, 2014) which adopted its current name in 1959 to strengthen its association with the motor vehicle company located there. However, cities have increasingly sought to work on their image through less drastic acts of reputation management and toponymic projects. A classic example is that of New York City where, amid the financial and social troubles of the 1970s, a coalition of elites (private and public) formed to craft a new urban image initiated a "coordinated, professional and global strategy" (Greenberg, 2008, p. 2). The now-universally recognisable abbreviation "NYC" is a downstream effect. This light-hearted naming tamed the city's image, helping to justify a range of other means of economic recovery. In a mere three letters, the wearer of a T-shirt can now be associated with the largest city in the United States and its pivotal place in global economic and cultural life. This link has long been recognised by property developers, tourism operators, and the cultural entrepreneurs of Broadway and high fashion, who convert the city's huge symbolic capital into financial capital (Morgan, Pritchard, & Pride, 2012).

Efforts to capitalize financially on the value of place names has introduced new elements to the multi-layered cultural politics of identity and naming associated with place, elements that have been increasingly understood as place branding (Anttiroiko, 2014). In Lury's (2004) terms, a brand is a set of relations between products that work to create sign-value through the "animated production of difference." Brands are constituted from (and as) a set of attributes and values, or equity, which Pike (2011b, p. 7) defines as "associations, awareness, loyalty, perceived quality and, crucially, origin." For Lash and Lury (2007), brands differentiate and create difference; they instantiate themselves in products but are inventive of wider social meaning, Fundamentally, they are part of post-commodity worlds and tied to economies of experience such as the global culture industry.

Pike et al. (2011a) demonstrate the many ways in which brands, and place brands in particular, are part of the geographical entanglement of commercial interests into everyday life. Place branding involves an attempt to re-engineer perceptions of communities, cities and nations either seen to be in need of regeneration or a boost in attempts to gain "an edge" in the competition to attract attention, tourists, arts, sporting events, and investment capital more broadly (Aronczyk, 2013). City-branding is a serious business that seeks to ensure that the reputational associations between certain places and goods are captured, become geographically entangled (Pike, 2011a) and put to work "in the name of" that place. At its heart lies the recognition that, in Medway and Warnaby's words, "a toponym can. . .deliver a degree of brand resonance in which the place name itself becomes the core of brand identity activities" (2014, p. 153).

The critical literature on place branding, however, tends to focus attention on civic boosterism, creative city initiatives, major events-based tourism, and economies of provenance. Further, it tends to confine its attention to larger metropolitan centres. With exceptions (Jackson et al., 2011; Light, 2014), it has not engaged in sustained critique of the entanglements of product branding, resistance to "toponymic capture" or the appropriation of place names as brands, along with the governance of place through branding. To a large extent, this is because – self-evidently – places are not corporations and lack both clear organisational boundaries and strategic centres. Meaningful and acceptable links between collective memory and the politics of urban space must therefore be established (Rose-Redwood, Alderman, & Azaryahu, 2008). To be successful place branding requires the active involvement and tacit consent of communities

(Eshuis & Edwards, 2013; Kavaratzis & Ashworth, 2008). In the case of a place name, in particular, "emotional ownership" can be tenacious and resilient. Resistance to change ought to be expected when a name has been *in place* long enough to have accrued strong associations with local identity and collective memory (Moon, Kearns, & Joseph, 2015).

Our paper addresses each of these gaps; it considers a place branding initiative at the regional scale that was entangled in economic development, premised on complete renaming and on governing place through the new name, and one that was successfully resisted by community actors. It examines the effort to rename the regional centre of Marlborough after the region's successful wine brand "Marlborough." Marlborough itself has been re-"imagineered" as a winemaking region over the last 35 years by a range of promotional agencies ranging from wine writers to tourism operators and formal regional and national government promotions agencies. Wines, landscapes, winemakers, soils and grapes have been carefully fashioned into a creative place-product relationship, expressed in the brand "Marlborough Sauvignon Blanc."

By way of precedent, there have been studies that have considered names and naming in the formation of new place identities as part of larger strategies of economic development (Urry & Larsen, 2011). For instance Rojek (1993) examined contrived names such as "Bronte Country" which, through allusion to literary tradition, essentially brand an entire region with reference to authors and fictional characters. Peter Jackson and colleagues (2011) have written about the creation of the Oakham chicken by Marks and Spencers in an effort to capture and sell quintessential Britishness embodied in a chicken, despite the fact that Oakham does not exist. In our study context, New Zealand cities have sought to create city brand values through campaigns such as Auckland's "City of Sails," Wellington's "Absolutely Positively Wellington," and Hamilton's ironic "More than you expect." Regional towns have associated themselves playfully with dominant agrifood icons (e.g., Ohakune as the home of the giant carrot; Te Puke's giant kiwifruit) which are designed to attract attention and tourists. The marketing of New Zealand as "Middle Earth" in homage to the Lord of the Rings film series both adds allure to tourism and disrupts a local sense of collective identity (Kearns, et al., 2015). To this extent, it is a short distance between marketing the country with a branded fantasy name and painting the fleet of the "flag-carrier" airline black (in allusion to the national rugby team, the All Blacks). Both are branding exercises that invoke, if not presume, a nation-wide "imagined community" (Anderson, 1983).

There is also a rich literature on provenance and the relationship between wine, reputation, branding and place. Wine is positioned in a social history and consumer culture dominated by notions of *terroir* (associations that link the qualities of wine to the place it was grown through soils, climate and winemaking traditions) (Moran, 1993). Place of origin matters in terms of cultural association, social status and economic value. As Moran (1993) has demonstrated, place-recognition has a crucial political economy, as illustrated by the French appellation legislation which controls production practice, land sales, and the marketing of wines according to strict product-place assumptions about quality. As a social product, a product branded by place and a product the quality of which is assumed to be determined by place, wine presents a rich palette for marketers seeking to turn association with place into references to quality and value.

The relationship is far from a marketing artefact, however, even if high volume and largely industrial wines might effectively be made in any place. As Harvey (2002) observes, the appellation system secures a rent generating relationship between place and product where toponymy is bound up with economic rent, monopoly powers, social status, and land prices. It supports monopolistic competition, but has been argued to impose rigidities on practice and to restrict innovation in production, marketing, and business practice. Beyond Europe, wine economy has been caught in a struggle between more artisanal producers who have sought similar appellation rules to support their wines in crowded markets and the globalising, free trade logics of larger scale producers who have sought greater flexibility in production technologies, geographies of production, and marketing. By and large countries have avoided the rigid rules of appellations, but supported individual wine enterprises and regional groupings to use place to create and mark exclusivity in other ways. In short, wine at all price points is marketed and sold through references to place. Taplin (2015), for instance, describes the efforts of Napa Valley winemakers to create a "cult" product and place-association through producing only small quantities sold through limited distribution channels; New Zealand has experienced a proliferation of similar regional and other sub-national place-branding initiatives that tend to entangle the brand values of specific places with those of Brand New Zealand (Lewis, 2015, Lewis and Le Heron, 2018). Our Marlborough case suggests another approach: an attempt to influence local identity in order to broaden place-association from wine-bottle labels and regional identity to the urban centre at its heart.

For our purposes, there are three important points to be drawn from this rise of place branding along with name association or replacement. *First*, as with other forms of contemporary toponymic commodification, place branding is commonly associated with neoliberalism (Eisenschitz, 2010). This is not just because neoliberalism is a defining transformative force of all recent social life but by the inscription of competition deeply into all forms of subjectivity, individual and collective. On the one hand, the neoliberal subject materialises as an entrepreneurial project of the self (for competitive places as much as competitive selves). Such projects and selves are the subjects and objects of branding (Klein, 2001). On the other hand, brands have given impetus to the search for and creation of differentiation that constitutes globalisation; the fabric of the global culture industry (Lash and Lury, 2007). They are in this way co-constitutive of the ideological and governmental projects of neoliberalism, as well as being entangled in the financial flows, global trade, and regime of globalising market governance and marketized social practice that we normally recognise as neoliberalism (Larner and Higgins, 2017). Neoliberalism offers an unavoidable interpretive framework for a Marlborough City proposal that is premised on competition among cities and winemakers, overwriting the past in the name of garnering attention in global marketplaces, and governing place in the interests of a mix of global and national capital freed from international trade restrictions (Lewis and Le Heron, 2018).

However, neoliberalism cannot be seen to be an over-determining explanation. As New Zealand critics as much as any other have stressed, neoliberalism takes the form of contingent, place-specific and often contradictory political projects that give it actualised forms (Larner, Le Heron, & Lewis, 2007). Even in New Zealand, which is widely cited as an exemplar of purist neoliberal reform that went further and faster in freeing up markets, marketizing the state, welcoming foreign capital, and resubjectifying citizens as rational, self-interested decision makers (see, for example Peet, 2012), neoliberalism has not simply been rolled out across space in an all-consuming wave (Larner, 2003; Lewis, 2012 2015). Rather, neoliberal reforms have coincided with the fashioning of a self-confident cultural nationalism. This condition has been interpreted as the throwing-off of a colonial "cultural cringe" as well as the institutionalisation of new state forms of commitments to the indigenous Māori language and the return of confiscated assets, all of which are at work in the replacement of colonial place names with the original Māori names for those places (Berg & Kearns, 1996). Explanation in the name of neoliberalism must account for this contingency and complexity, and must also recognise the presence of other political projects and effective resistance.

Second, as Pike et al. (2011a) recognise, the branding of places can promote the appropriation of traditional meanings by branders and their corporate clients, distort identities, and foster new inequalities of access and patterns of exclusion. These "entanglements" of branding efforts can lead to new and unexpected spatial associations and connotations (Pike, 2011). New coalitions of branders and economic leaders in effect assume the right to attach meanings to cities and their names, if not to rename them *per se*. Place branding is therefore, as with naming in general, far from an innocent practice. *Third*, in the case of Blenheim, the normalizing of smaller-scale place-branding has provided inspiration for an audacious attempt to actually rename an urban centre as a brand.

Naming and claiming place in New Zealand

New Zealand is a white settler nation-state founded on colonial practices of naming and claiming. Its colonial toponymic landscape represents the systematic erasure of many indigenous meanings of place and all the socio-spatial processes of identity making and sovereignty associated with them. There can be no clearer demonstration of the depth of the politics of naming than this widespread erasure. The legacy of this erasure remains prominent in the on-going politics of indigeneity, redress, and Māori sovereignty that now define national presents and futures. Indeed, some of this wider politics is being worked out through debate about place names themselves. For example, the use of "Aotearoa," the Māori name for the country, has increased in recent years, usually in close association with the enduring European name bestowed by Dutch explorer Abel Janzoon Tasman in the seventeenth century. The commitments made to *te reo* (Māori language) in the Māori Language Act of 1987 and to Māori language education are reflected in the increasing use, and improved pronunciation, of Māori place names in public spheres such as broadcasting (Kearns & Berg, 2002).

This political post-colonial project of language and naming is also reflected in the work of the New Zealand Geographic Board, the statutory authority responsible for official place names. The majority of its work involves dealing with the often politically contentious process of adjudicating upon and restoring Māori place names. This process can involve contest between the competing claims of different Māori groups to place and place names as well as between Māori and settler communities that have grown up with colonial names (Stokes, 1994). Through its work, renaming places in New Zealand has become strongly associated with a positive progressive politics (Berg & Kearns, 1996).

Running in parallel with this restitutive project of renaming places is a rampant economic nationalism. In his account of the building and work of "Brand New Zealand," Lewis (2015) points to how it has galvanised the re-imagination of nation as corporation – "New Zealand Inc" – in which national and subnational identities, cultural practices and individual and collective energies are to be yoked to export growth. The nation is positioned as the primary collective socio-cultural and economic identity, and the mobilisation of economic effort at the national scale is seen as essential to ensure economic success as a small economy distant from its markets. It is also positioned as brand identity itself, "100% Pure *New Zealand*." While other nations have developed a nation brand, the notion of "Brand New Zealand" is especially deeply ingrained in the national psyche of a country that seeks to make itself seen and recognised from a distance in an increasingly globalised world (Bell, 1996). It is potentially unique in the extent to which it is attached in carefully orchestrated ways to the nation understood as corporation.

The national referendum over a proposed new flag (2015/16) is a case in point and one that indicates the social power at work in "Brand New Zealand" and New Zealand Inc. The campaign to change the flag was led by a coalition of the new wealthy represented by the corporate leaders of New Zealand Inc., including Prime Minister John Key (Osborne, Lees-Marshment, & Van Der Linden, 2016). It was orchestrated by the architects of Brand New Zealand (Lewis, 2016) conducted by its proponents in the language of branding as a search for a flag that would better represent the national brand. These interests made bold claims to the right to rebrand the nation and, to the extent that name and brand are increasingly co-articulated in the production and representation of place, to in effect rename it. In the next section, we examine how this project has translated into a proposed renaming of a "city" in the image of a place-brand with the intention of capturing its reputational benefits.

Blenheim: from Bavaria to Aotearoa

Located on the north of New Zealand's South Island, or Te Waipounamu, Blenheim is situated at the base of the extensive Wairau alluvial plain between two topographic features: the Wither Hills and the Richmond Ranges. It is one of New Zealand's sunniest climates with hot and largely dry summers and frosty winters. Its soils are characterized by nutrient-poor river gravels. As a colonial settlement, Blenheim was originally known by working settlers as "Beaverton" in reference to its flood-prone character (Lewis, 2016a). The settlement was developed in the mid-nineteenth century as an administrative and service centre for the Marlborough region which, in well-rehearsed colonial style, was named to honour a British aristocrat, John Churchill, the first Duke of Marlborough (Reed, 2002). Blenheim was named for the victory of his armies over the French in 1704 at the Bavarian village of Blindheim (Anglicized as Blenheim). The name displaced the original Māori name for the place, Waiharakeke or "the waters of flax."

Over the last 35 years, the Wairau plains have been transformed from poor quality, low-yielding sheep pasture to high-value vineyards that make globally distinctive Sauvignon Blanc wines. Blenheim has changed from a small regional service town to a recognised centre of the global wine industry. It is home to 65% of New Zealand's land planted in grapevines and source of over 80% of its wine exports. Blenheim's population grew from less than 18,000 in 1981 to nearly 31,000 in 2016, while the number of hectares under grapes in the region increased by over 1,000% during the same period. The "city" and wine industry have grown together such that Blenheim is deeply connected with wine and the valley is a monocultural vinescape.

The relationship between town and industry is rooted in a mix of microclimate, alluvial gravels, grape clone selection, historical accident, and strategic human practice in the vineyard, winery, marketing department and boardroom that gives Marlborough Sauvignon Blanc its global distinctiveness (Moran, 2016). The relationship was forged in the early 1970s when Montana wines discovered a series of relatively cheap land parcels on the plains in the midst of a struggle with its competitors to dominate the domestic wine market. The valley was, at the time, seen as low quality land best suited to sheep farming. While Montana initially struggled to make the vines grow due to lack of water, the combination of environmental, geophysical and human factors at work in the valley imparted highly distinctive aroma profiles to Sauvignon Blanc when the grapes were thresh-harvested in industrial quantities (Trought & Bramley, 2011). Marlborough Sauvignon Blanc has proven to be one of the world's most distinctive and profitable large-volume varieties, helping to informally identify this region as synonymous with high-quality wine as have other regions like the Barossa (Australia), Okanagan (Canada) and Napa (US) valleys (Getz & Brown, 2006). Yet our concern is not the "winescape" (Hall & Prayag, 2017), but rather how particular viticultural interests attempted to amplify the identity of the Marborough wine region through transforming the name of Blenheim.

Marlborough's growth as a wine producer coincided with the growth of wine tourism, adding a second layer of significance to associations between product and place. Driven by the interacting allure of the rural idyll, artisanship, alcohol and landscape aesthetics, wine tourism added new impetus to the growth of wine regions in many parts of the world. Strong associations between place, landscape and product now offered new opportunities to create economic value. "Wine branding by place" became joined by "place branding by wine" as a target of economic interest for wine companies, regional development agencies and a host of connected marketing and branding experts. Blenheim was no different. As the industry in Marlborough has consolidated over the last two decades, production has increasingly been complemented by tourism with vineyards offering restaurants, tasting rooms and tours, and selling all manner of cellar-door merchandise. Promoted as "Marlborough – Wine Country," the region is now also home to a sold-out annual "Marlborough Wine and Food Festival." Tourism has expanded and has connected a wider set of interests into the values associated with the branding and sale of place. It has cemented "Marlborough" as a distinctive place, tourist destination, and value-creating proposition for a greater range of people.

The naming and claiming of Blenheim

> I will loose (sic) my identity along with everyone else who was born there. There are a lot more important things to worry about – traffic flow, taking care of the sick, housing everyone so they are warm and sheltered (correspondent, *The Marlborough Express*, 25.5.16).

In May 2016, a group of business-people associated with the wine and hospitality industries in Blenheim announced through the local print media a proposal to change Blenheim's name to Marlborough City. This suggestion was reportedly made on account of the regional name having "more prestige than Blenheim" (Lewis, 2016a). The ensuing debate was covered in the regional and national media, with locally based reporters interviewing the leading protagonists and reporting their views. We draw on a systematic web-based review of all news articles pertaining to this proposal for the two months following the proposal's announcement in May 2016. This search yielded seven articles on nationally-available news-sites amounting to 2,793 words of text and 12 minutes of radio interviews. In what follows we draw on this media coverage with particular attention to the on-line comment opportunity that follows many items in New Zealand newspapers and in particular *The Marlborough Express*. Examination of selected comments allows us access to a form of discursive engagement with a controversial issue and expression of identity politics that, until recently, was only played out explicitly by fewer correspondents in the more formal letters to the editor page of a newspaper and also more ephemerally on talkback radio. We contend that jotted comments in response to the trigger of an online article offer a window into people's views that would not necessarily be as available in the context of a formal interview. Further, their examination allows us to "follow" (see Cook, 2004) this debate into the community in a way not easily achieved as an "outsider" academic interviewer bound by imposing the formalities of ethical protocols.

A key advocate for the name change was Mitchell Gardiner, general sales manager of the Allan Scott Family Estate, a pioneering independent winemaking business in the region. Gardiner's argument was that "when you travel internationally the recognition is definitely with Marlborough, it's certainly not with the name Blenheim" (Satherley, 2016). According to Gardiner, "the town is missing out on visitor dollars because of a lack of name recognition" (Lewis, 2016a).[2]

Some suggested "Blenheim" to be boring, outdated and unmarketable and deserving replacement. Others sought to distance the town's name from its association with "Blenheimer," a mass-consumption brand of budget-priced, boxed wine named after the town. Josh Scott, founder of the local, successful and brand-conscious Moa Brewing Company and son of Allan Scott, spoke to this concern, saying that a name change "will at least end the town's association with Blenheimer, a pretty horrible wine" (Satherley, 2016). This view was supported by a correspondent to *The Marlborough Express* who noted, "Anyone who has memories of Blenheimer, a nasty, sweet budget wine, will agree to the name change, for sure."

A further imperative was broader economic development. According to Scott, in a thinly-veiled reference to the extraction of economic returns, "we can get more value out of Blenheim just by making a small change to Marlborough City." Lewis (2016b) reports the proposal's backers as suggesting that the name change might "help rejuvenate the Blenheim town centre" with an increase in visitor numbers being envisaged as helping to boost the retail and hospitality sectors. Reference to city-status in the proposed replacement name, despite Blenheim's modest population, did not deter the proponents. "Technically we're not a city," said Scott, "but we're not really getting stuck up on the semantics, it's about giving us more of a profile" (Lewis, 2016a).

The marketing imperative was clearly evident in Gardiner's comment that international visitors "might know what Marlborough is or particularly why they want to visit Marlborough, but they can easily drive through Marlborough without even realising they've been here." Hence the proposal was as much about giving *more* prominence to the Marlborough name as it was a quest to *replace* the name Blenheim. However, ironically perhaps, Gardiner is reported to have developed the idea of changing Blenheim's name after being at a stand-up performance by Dai Henwood, a popular New Zealand comedian, who "mercilessly mocked the town" for its blandness (Satherley, 2016).

Moa Brewery's Josh Scott lent support to the idea of a name change: "Marlborough is such a great marketing opportunity, and it's a way that we can jump on board and help promote the whole region – not just the wine industry but also the Sounds, we've got some great walks and mountain bike parks." Thus, an attempt was being made to engineer a name change so as to align the same – already well-known – name to both the region and its service centre.

The proposal's aftermath

Classic marketing ploy, no need to improve the product, just sell it in a fancy wrapping (correspondent, *The Marlborough Express*, 25 May, 2016)

Reaction to this attempt to leverage off the success of the wine industry and rebrand the town was swift. Perhaps predictably, the Mayor of Marlborough District, Alistair Sowman, was cautious about the idea. The mayor was described as "sitting on the fence," calling the proposal "an interesting idea," but unsure of how widespread support would be, especially given Blenheim falls below the current numeric threshold for city status. Further, according to Sowman, "there's a lot of history here" and "Marlborough is about more than the wine industry, with about 20 per cent of New Zealand's coastline … and a growing aquaculture industry" (Lewis, 2016a).

The proposal led to a rapid flurry of commentary and national news coverage. In the comments section of *The Marlborough Express*, one reader proclaimed surprise that "they haven't decided to change the name to that of a wine company" in clear recognition of this proposal's subtext as a marketing move. A second correspondent wrote, "A home for many people should not become an advertiser's tool" and another, unwittingly suggesting the geographical entanglements idea (Pike, 2011b), said the idea was guilty of "conflating a 'brand' with a proper noun and place name."

The potential slippage between being known for wine and confusion with tobacco was noted by a correspondent: "Marlborough City just sounds like the cigarette capital of the world."[3] Another signalled the proposal as a potential threat to residents' identities and that this issue was a distraction to more pressing concerns. In an appeal to birthright, another wrote, "Perhaps only those born in Blenheim should make this decision?" This view was supported by local historian, John Orchard, who doubted the proposal would gain any traction because, "You've got around 150 years of the biggest urban area in Marlborough being called Blenheim" (Lewis, 2016a). In a similar vein, Glenn Howard, the founder of a Facebook page, *Old Marlborough, the way we were*, said that its 10,500 members were firmly against the idea. According to Howard, "The name Blenheim is etched into the history of Marlborough" (Lewis, 2016a).

Other correspondents saw more effective ways to market the region than changing the town's name. These included, "A decent 5 star Hotel would do more for Marlborough than a name change" and "A name is not going to fix it and will cost lots of money that could be better spent on improving attractions." The blatant ambition of proposing that a town be promoted to city status was noted: "If you want to attract tourists develop the area, calling a town a city would only generate further disappointment."

Along with a minority of commentators, Blenheim-based Member of Parliament and former Chairman of the New Zealand Winegrowers Stuart Smith offered a different view. He noted that "if we are going to think about a name change, we need to think about a name that reflects our heritage and consideration could be given to a Māori name – such as Blenheim's original name, Waiharakeke (*The Marlborough Express*, 2016). For reasons discussed below, this Māori name had been largely written out of the conversation of May 2016, unlike elsewhere in New Zealand where, for instance, the restoration of indigenous names to the landscape is increasingly being recommended by the statutory body responsible for adjudicating place name disputes, the New Zealand Geographic Board. For one commentator, "Waiharakeke has a nice ring to it! This was the original name so why not return?" and another "Waiharakeke 'the waters of flax' should be reinstated as it was for six hundred years before foreign investors arrived!"

Interpreting 'Marlborough City'

> *"If the people in the wine industry want to do something for the region, then they should put better labels on their wine and start marketing Blenheim."* (Local resident, cited in Lewis, 2016c)

Why, after over 150 years of comfortable acquiescence to the name of Blenheim, were a group of its citizens led to contemplate changing its name, and why did they fail? In her book, *Inventing New Zealand*, Bell (1996, p. 1) notes, "New Zealand is a country that constantly tells itself who it is. Every single day there are messages and statements to affirm that this is a great country." In our view the proposed renaming of Blenheim was akin to a corporate "takeover bid" that sought to rename a city as a way of better managing the symbolic landscape in the cause of economic growth and a more marketable local economy and branded place (see Semian, Chromý, & Kučera, 2017). It sought to convert the cultural capital of recognition into financial gain. The proposed new name of Marlborough City potentially offered two paths of conversion: first, in its use of "City," it created a narrative of cosmopolitan urbanity for the people of Blenheim and surroundings; and second, through highlighting Marlborough, it erected a toponymic billboard affirming the wine region's greatness to domestic and international markets.

Drilling down into this claim, we identify three key drivers underlying the takeover bid, each of which can be linked to the context of neoliberal ideology. *First*, the initiative can be connected with the dominance of economic identity over socio-historical identity at both the local and national scale, and thus the rise to prominence of economic objects and rationalities of government over all others. This prominence can, in turn, be associated with the dominance of neoliberal ideology in New Zealand society at large. At the level of national identity, this prominence has translated into what has been called economic nationalism (Lewis et al., 2008): the idea that national identity is wrapped up in economic rationalities, performance and objectives. In New Zealand, this ideology is reflected in the reduction of national interest to a concerted drive to generate exports by harnessing resources and human energies to the demands and bottom lines of what could be seen as a mythical "NZ Incorporated."

The notion of a national, collective corporate identity or mentality comes with a series of "Brands New Zealand" whereby cultural and environmental resources, legacies and imageries are connected to particular products and to an overall Brand New Zealand which can be attached to exported products. The ultimate goal is to draw attention to them in crowded global market places. Arguably, only in such a neoliberal environment could a town be renamed so as to align itself with its region's primary product, the Savignon Blanc, a globally distinctive wine. Further, only in a neoliberal era, could we imagine the subjects who might drive such a project as audacious as proposing to rename a town of over 30,000; subjects thoroughly imbued with market ideology, economic identities of place, and the reduction of the social to an economic rationality.

Second, the name change proponents were broadly representative of the same new wealthy social group who sought, only months earlier, to replace the national flag with symbols that purportedly better reflected the nation's brand (Osborne et al., 2016). The flag change example has direct parallels with renaming Blenheim – all social identities are up for grabs in the project to increase profit and rebrand New Zealand in the interests of those social groups who do the branding and make the products behind the brands. In this case, the group seeking to rename Blenheim was already a major beneficiary of neoliberal free trade; an industry that owes much of its success to free trade, the freeing up of liquor retailing, and foreign investment in New Zealand (Hayward and Lewis, 2008).

Third, the new wealthy who sought to rename Blenheim are among the children of neoliberalism who Nairn, Higgins, and Sligo (2012) warned would be its most dangerous legacy. Now into their mid-thirties and early forties, they have been brought up neoliberal and owe much of their wealth and socio-political status to a neoliberalism that makes renaming place in order to do business more effectively both thinkable and doable. In what other circumstances might it even be thought appropriate to attempt to overwrite both heritage and, perhaps unintentionally, indigenous aspirations? Our case study thus serves as a potent example of Light and Young's (2015) assertion that, in certain contexts, place names have as much an economic as political role in promoting ideological interests.

The "Marlborough City" proposal was an audacious campaign, and one that exerted – but ultimately failed – to harness sufficient political power to establish legitimacy. Its momentum began and ended with radio interviews and print media investigations of the merits and pitfalls of a change. While discussed by the mayor and Member of Parliament, the proposal lost momentum before any attempt to formalise a proposal through, for example, petitioning of the New Zealand Geographic Board. In underestimating the limits of economic nationalism as a political discourse, the significance of place meanings and the associated intensity of toponymic politics, the proposal failed. Within two weeks the story and discussion disappeared from national news media coverage. It encountered a range of interests, including representatives of the Blenheim "establishment" for whom heritage mattered and Māori who saw the proposal as a lost opportunity to reconsider use of the town's original name. Like the efforts to change the national flag, it failed. In this case there were too few Blenheim citizens involved for the proposal even to gain the support that would be needed for formal consideration (by the New Zealand Geographic Board).

The Marlborough City case aligns with Bell's observation that such initiatives tend to be the domain of the dominant Pākeha (European New Zealander) majority who "less than two centuries after settlement [are still] .... developing landmarks to solidify notions of identity" (1996, p. 29). Some landmarks are symbolic, in this case claiming the right to name and rename under the banner of economic competitiveness rather than colonial might. Yet, this project is contested. On the one hand, the name Waiharakeke endures and is still spoken "under the radar" of everyday discourse. It's persistence speaks to the way even a rarely uttered name can endure in memory, notwithstanding colonial actions that have led to the physical and symbolic alienation of first peoples (Kearney & Bradley, 2009). On the other hand, the decolonisation project led by Māori has yielded a gathering popular and institutional momentum for a different project of re-inscribing physical features and places with their original Māori names (Kearns & Berg, 2002). Both these expressions of the pursuit of a new symbolic order are at work in the resistance to "Marlborough City" and are articulated in complicated nuanced ways.

Conclusion

Our case study amounts to what we interpret to be an attempt to galvanise the "placeness" (Molotch, 2002) of the Marlborough brand and engage in a politics of "re-imagining" place (Johansson, 2012). Yet city branding involves more than simply image; it is invariably political and a means by which citizens can gain or lose influence on their collective future (Eshuis & Edwards, 2013). The case for the proposal to rename Blenheim was neither strong enough nor concerted enough to overcome opposition to it. Neither the problem it offered to fix nor the benefits it promised to provide were seen as particularly persuasive, even in its own terms of New Zealand Inc. However, there was something more significant going on in the proposal than an opportunistic, somewhat less-than-full-blooded move aimed at enhancing commercial returns. It is to this deeper politics that we turn to conclude.

Place names are increasingly deployed as brands in strategic re-imaginings of their role or symbolic worth. As with all brands (Lury, 2004), they are replete with possibility yet also reset boundaries of inclusion and exclusion (Johansson, 2013). For a brief moment in 2016, urban governance in the Marlborough region was destabilised by the potential persuasiveness of a proposal to alter the symbolic identity of its urban settlement. That Blenheim did not succumb speaks to the politics of inclusion and exclusion at stake in the renaming of an entire town. The fate of the proposal demonstrates that there is a considerable gulf between achievable goals such as gaining naming rights to a sports stadium (Light & Young, 2015) and that of an entire town. Legitimacy involves aligning interests and the conversion of power into authority (Eshuis & Edwards, 2013). In this case study, a selection of Marlborough region's winemakers were unable to align support and convert their economic power into political leverage.

The question we posed when we drew on a critique of place branding to interpret the "Marlborough City" proposal was whether the campaign reached beyond superficial rebranding and embraced claims-making in the politics of toponymy. Our answer is that the case clearly illustrates the exercise of power and exclusion in relation to the capacity to rename the city as brand, to rewrite histories of place, and to plot new futures. The proposal was a far from innocent act, and reflects the rise of branding and the intensifying commodification of place and tradition through cultural economy under neoliberalized post-modernity. It also illustrates some of the deeper, transformative politics of neoliberalism in the form of the appeal to rationalities and the performance of subjects fostered by the neoliberal project for over thirty years. "Marlborough City" was offered up by its proponents as simply more common sense; it was cast as another expression of the surrender of all collective identity to Brand NZ and "New Zealand Inc." To the extent that it was short-lived, the proposal speaks not just to the prominence of that common sense but also to the proposal's arrogance; there was not enough to be gained to fight hard, but enough of a disregard for others (past, present and future) born of economic superiority to try in the first place.

We conclude that this attempt to alter the toponymic landscape, while imagined as economically creative, was at the same time an act of exclusion and erasure of social histories by a new economic occupation of the region. In the first instance, New Zealand Inc. is, in its own terms, contradictory. The Marlborough City proposal is illustrative of a defiantly post-colonial yet recolonising New Zealandness. The emotion underlying the self-confidence of the idea speaks to the way branding is "not particularly aimed at either deliberation or reason" (Eshuis & Edwards, 2013, p. 1068). This condition of new-New Zealandness is, of course, inextricable from its proponents having grown up alongside a new found economic confidence and capability to extend a national reach well beyond proportionality for a topographically isolated country of 4.5 million. But it has come with an ability to "punch above one's weight" and to be winning international awards for music, film, olives and wine. The net result is that at the level of nation, local history and regional culture is reduced to a resource to be mined and mobilised for purpose of branding and creating wealth for its beneficiaries. There are deep contradictions here to which the politics of naming and claiming points, and does so ahead of the emergence of these contradictions in other political fields.

The proposal amounted to an implicit attempt to silence indigenous Māori toponymy in favour of the neoliberal ideal of a name as a marketing opportunity (Vuolteenaho & Berg, 2009). It failed, in large part because that toponymy, a community that embraced it, and the institutionally resourced indigenous renaming project would not be silenced. If naming is a form of norming (Berg & Kearns, 1996) then this resistance amounts to more than simply a dislike for "Marlborough City" or a preference for Waiharakeke. Rather it is both a rejection of the symbolic order of naming and material order they create – one which offers legitimacy to those whose neoliberalised ethos has allowed a dominance of the politics of place representation. It is a challenge to the discourse of New Zealand Inc. and New Zealand's neoliberalism more generally. The irony is that both Blenheim and the regional name of Marlborough are colonial names, such that any change would simply create further distance from the original Māori, culturally "proud," defiantly post-colonial, New Zealand name of Waiharakeke, which had been in place for centuries prior to European settlement.

This example of the "entanglements of language with power relations" (Vuolteenaho & Berg, 2009, p. 1) is deeply material. Representatives of business interests have sought to fly a flag for their land-based enterprises by creating a newly neoliberalised land-mark: a rebranded name for the urban place that serves the region. That this attempt to change a name potentially affected the 30,000-plus residents speaks to the agency of those with economic leverage. Two months after the proposal, references to it in the local media had disappeared and a year later, the proposal to change Blenheim's name appeared to have lost any currency, speaking to the agency of dissenters. In fact, it was the strategic intent and co-option of the name that appears to have led to most of the rapid resistance. For residents of the region, renaming their town was a step too far.

In recounting this story, we have implicitly addressed Rose-Redwood, Alderman, & Azaryahu's (2010, p. 466) call "to consider the social costs of indiscriminately allowing the *exchange-value* of a name to triumph over its *use-value* in the public sphere." To close, we contend that even if, as Josh Scott observed, the proposal's failure was still a useful exercise in marketing and had "got people thinking about how great the region was, and how Blenheim should be marketed" (Lewis, 2016c), it was clearly an overreach by the interests of a particular social group in what remains a politically contested present. That this proposal to market Marlborough by changing the name of its urban centre failed means that ours is an account of a local political contest writ through with the resilience of place and heritage.

Notes

1. Ironically, perhaps, a change in the threshold to meet city–status to 50,000 in the 1989 reforms to local government meant that Blenheim no longer holds city-status.
2. Blenheim is not the first New Zealand urban centre to sustain attempts at a name change. The nation's fourth-largest city, Hamilton, shares its name with numerous other places across the British Commonwealth and has toyed with name changes to "Waikato City" (after its surrounding region) and "Fountain City." Both ideas were widely ridiculed. There is also the case of Palmerston North, a regional New Zealand city named after Lord Palmerston, who was a British Prime Minister in the mid-nineteenth century. Then Deputy Mayor, Jim Jeffries, led a 2013 campaign and petition to the New Zealand Geographic Board, for the city to be renamed Manawatu City, after the surrounding region (Hyde 2013a; 2013b).
3. There are layers of irony behind this remark. The first grapes were planted in Marlborough by Montana wines, which has remained the region's most prominent and largest wine producer, albeit now as a subsidiary of Pernod Ricard. Its wine is rebranded as Brancott Estate, taking up the name used to brand their wines in the USA where they were prohibited from using the name Montana. In 2002, Montana bought out Corbans wines, New Zealand's second largest wine producer at the time which in turn had been bought by Rothmans in 1979. Proposing to rename their town Marlborough, a name in another

context associated with the now beleaguered tobacco brand, appears to be a bizarre latest move in the politics of branding in this place.

Disclosure statement

No potential conflict of interest was reported by the authors.

ORCID

Robin A. Kearns http://orcid.org/0000-0002-7890-1218
Nicolas Lewis http://orcid.org/0000-0001-8645-0704

References

Anderson, Benedict. (1983). *Imagined communities: Reflections on the origin and spread of nationalism*. London: Verso.

Anttiroiko, Ari-Veikko. (2014). *The political economy of city branding*. London: Routledge.

Aronczyk, Melissa. (2013). *Branding the nation: The global business of national identity*. London: OUP.

Bell, Claudia. (1996). *Inventing New Zealand: Everyday myths of pakeha identity*. Auckland: Penguin.

Berg, Lawrence D., & Kearns, Robin A. (1996). Naming as norming? 'Race', gender and the identity politics of naming places in Aotearoa/New Zealand. *Environment and Planning D: Society and Space*, *14*(1), 99–122.

Cook, Ian. (2004). Follow the thing. *Antipode*, *36*(4), 642–664.

Eisenschitz, Aram. (2010). Place marketing as politics: The limits of neoliberalism. In Frank. M Go & Robert Govers (Eds.), *International place branding yearbook 2010: Place branding in the new age of innovation* (pp. 21–30). New York: Springer.

Eshuis, Jasper, & Edwards, Arthur. (2013). Branding the city: The democratic legitimacy of a new mode of governance. *Urban Studies*, *50*(5), 1066–1082.

Evans, Graeme. (2003). Hard-branding the cultural city- from Prado to Prada. *International Journal of Urban and Regional Research*, *27*(2), 417–440.

Getz, Donald, & Brown, Graham. (2006). Benchmarking wine tourism development: The case of Okanagan Valley, British Columbia, Canada. *International Journal of Wine Marketing*, *18*(2), 78–97.

Greenberg, Miriam. (2008). *Branding New York: How a city in crisis was sold to the world*. New York: Routledge.

Hall, Colin Michael, & Prayag, Gilish. (2017). Guest editorial. *International Journal of Wine Marketing*, *29*(4), 338–345.

Harvey, David. (2002). The art of rent: globalization, monopoly and the commodification of culture. *Socialist Register*, *38*, 93-110.

Hayward, David, & Lewis, Nicolas. (2008). Regional dynamics in the globalising wine industry: The case of Marlborough, New Zealand. *The Geographical Journal*, *174*(2), 124–137.

Hyde, Chris. (2013a, May 9). Deputy mayor wants Palmerston North name change. *Manawatu Standard*. Retrieved from www.stuff.co.nz/manawatu-standard/

Hyde, Chris. (2013b, May 11). Mayor doubts name change worth it. Manawatu Standard. Retrieved from www.stuff.co.nz/manawatu-standard/

Jackson, Peter, Russell, Polly, & Ward, Neil. (2011). *Brands in the making: a life history approach*. In Pike, Andy Brands and Branding Geographies (59-74). Cheltenhamn UK: Edward Elgar.

Johansson, Marjana. (2012). Place-branding and the imaginary: the politics of re-imagining a garden city. *Urban Studies*, *49*(16), 3611-3626. doi:10.1177/0042098012446991

Johansson, Marjana. (2013). Place branding and the imaginary: The politics of re-imagining a Garden City. *Urban Studies, 49*(16), 3611–3626.

Kavaratzis, Mihalis, & Ashworth, Gregory. (2008). Place marketing: How did we get here and where are we going? *Journal of Place Management and Development, 1*(2), 150–165.

Kearney, Amanda, & Bradley, John J. (2009). 'Too strong to ever not be there': Place names and emotional geographies. *Social and Cultural Geography, 10*(1), 77–79.

Kearns, Robin A, & Berg, Lawrence D. (2002). Proclaiming place: Towards a geography of place name pronunciation. *Social and Cultural Geography, 3*(3), 283–302.

Kearns, Robin, O'Brien, Gregory, Foley, Ronan, & Regan, Nell. (2015). Four windows into geography and imagination(s). *New Zealand Geographer, 71*(3), 159–176.

Klein, Naomi. (2001). *No logo: no space, no choice, no jobs*. London: Flamingo.

Larner, Wendy. (2003). Neoliberalism? *Environment and Planning D: Society and Space, 21*(5), 509–512.

Larner, Wendy and Higgins, Vaughan (Ed.s) (2017). *Assembling neoliberalism: expertise, subjects, practice*. New York: Springer.

Larner, Wendy, Le Heron, Richard, & Lewis, Nicolas. (2007). Co-constituting 'after neoliberalism': Globalising governmentalities and political projects in Aotearoa New Zealand. In Kim England & Kevin Ward (Eds.), *Neoliberalization: States, networks, peoples* (pp. 223–247). New York: Wiley.

Lash, Scott, & Lury, Celia. (2007). *Global culture industry: the mediation of things*. Malden, MA: Polity Press.

Lewis, Nicolas. (2012). Splitting a northern account of New Zealand's neoliberalism. *New Zealand Geographer, 68*(3), 168-174. doi:10.1111/nzg.2012.68.issue-3

Lewis, Nicolas. (2015, June 29). Is our flag a brand? Sciblogs, Science Media Centre, Royal Society of New Zealand. Retrieved from https://sciblogs.co.nz/

Lewis, Nicolas. (2011). Packaging political projects in geographical imaginaries: The rise of nation branding. In Andrew Pike (Ed.), *Brands and branding Geographies* (pp. 264–287). Cheltenham: Edward Elgar.

Lewis, Nicolas, Larner, Wendy, & Le Heron, R L. (2008). The New Zealand designer fashion industry: making industries and co-constituting political projectsa. *Transactions Of The Institute Of British Geographers, 33*(1), 42-59. doi:10.1111/j.1475-5661.2007.00274.x

Lewis, Oliver. (2016a, May 23). Should Blenheim be renamed Marlborough City as nod to wine industry? Retrieved from www.stuff.co.nz

Lewis, Oliver. (2016b, May 24). Proposal to change Blenheim's name to Marlborough City not that unusual. Retrieved from www.stuff.co.nz/marlborough-express/

Lewis, Oliver. (2016c, May 25). Marlborough City debate finds common ground in need to consider Blenheim's marketing. Retrieved from www.stuff.co.nz/marlborough-express/

Light, Duncan. (2014). Tourism and toponymy: Commodifying and consuming place names. *Tourism Geographies, 16*(1), 141–156.

Light, Duncan, & Young, Craig. (2015). Toponomy as commodity: Exploring the economic dimensions of Urban place naming. *International Journal of Urban and Regional Research, 39* (3), 435–450.

Lury, Celia. (2004). *Brands: The logos of the global economy*. London: Taylor and Francis.

Medway, Dominic, & Warnaby, Gary. (2014). What's in a Name? Place branding and toponymic commodification. *Environment and Planning A, 46*(1), 153–167.

Molotch, Harvey. (2002). Place in product. *International Journal of Urban and Regional Research, 26*(4), 665–688.

Moon, Graham, Kearns, Robin, A, & Joseph, Alun E. (2015). *The afterlives of the psychiatric asylum: Recycling concepts, sites and memories*. Farnham: Ashgate Press.

Moran, Warren. (1993). The wine appellation as territory in France and California. *Annals of the Association of American Geographers, 83*(4), 694–717.

Moran, Warren. (2016). *New Zealand wine: The land, the vines, the people*. Auckland: Auckland University Press.

Morgan, Nigel, Pritchard, Annette, & Pride, Roger (Eds.). (2012). *Destination brands: Managing place reputation*. London & New York: Routledge.
Nairn, Karen, Higgins, Jane, & Sligo, Judith. (2012). *Children of Rogernomics: A neoliberal generation leaves school*. Dunedin: Otago University Press.
Nicolas, Lewis, & Le Heron, Richard (2018). New Zealand wine: Seeking success beyond growth. In Pawson, Eric (Ed) *The New Biological Economy* (116-136) Auckland University Press: Auckland.
Osborne, Danny, Lees-Marshment, Jennifer, & van der Linden, Clifton. (2016). National identity and the flag change referendum: Examining the latent profiles underlying New Zealanders' flag change support. *New Zealand Sociology*, *31*(4), 19–47.
Peet, J. Richard. (2012). Comparative policy analysis: Neoliberalizing New Zealand. *New Zealand Geographer*, *68*(3), 151–167.
Pike, Andy (Ed.). (2011a). *Brands and branding geographies*. Cheltenham: Edward Elgar.
Pike, Andy. (2011b). Placing brands and branding: A socio-spatial biography of Newcastle Brown Ale. *Transactions, Institute of British Geographers NS*, *36*, 206–222.
Reed, A. W. (2002). *The reed dictionary of New Zealand place names*. Auckland: Reed Books.
Rojek, Chris. (1993). *Ways of escape*. Basingstoke: Macmillan.
Rose-Redwood, Reuben, Alderman, Derek, & Azaryahu, Maoz. (2008). Collective memory and the politics of urban space: An introduction. *GeoJournal*, *73*(3), 161–164.
Rose-Redwood, Reuben, Alderman, Derek, & Azaryahu, Maoz. (2010). Geographies of toponymic inscription: New directions in critical place-name studies. *Progress in Human Geography*, *34*(4), 453–470.
Satherley, Dan (2016). Winemakers want to rename Blenheim 'Marlborough city'. Newshub 24th May (www.newshub.co.nz).
Semian, Michael, Chromý, Pavel, & Kučera, Zdenek. (2017). Name as regional brand. *Journal of Language and Politics*, *15*(6), 768–789.
Stokes, Evelyn. (1994). Geographic naming in New Zealand: The role of the New Zealand geographic board. *New Zealand Geographer*, *50*(2), 51–54.
Taplin, Ian M. (2015). Crafting an iconic wine: The rise of cult Napa. *International Journal of Wine Marketing*, *28*(2), 105–119.
Trought, Mike. C. T., & Bramley, Robert. G. V. (2011). Vineyard variability in Marlborough, New Zealand: Characterising spatial and temporal changes in fruit composition and juice quality in the vineyard. *Australasian Journal of Grape and Wine Research*, *17*(1), 79–89.
Urry, John, & Larsen, Jonas. (2011). *The tourist gaze 3.0*. London: Sage.
Vuolteenaho, Jani, & Berg, Lawrence D. (2009). Towards critical toponymies. In Berg, Lawrence and Vuolteenaho, Jani (Ed.s), *Critical toponymies: The contested politics of place naming* (pp. 1–18). Farnham: Ashgate.

Afterword

The names of urban dispossession: a concluding commentary

David J. Madden

ABSTRACT

This concluding commentary discusses some of the ways in which branded renaming reflects conflicts over ownership, authority, property and power. Practices of sponsored renaming and place branding are not monolithic, but this special issue shows that they are deeply rooted within neoliberal urbanism. It should therefore not be surprising that place branding is almost always contested, and often a flashpoint in struggles over urban space.

The neoliberal city is instantly identifiable simply by looking at a map. Contemporary London, for example, is home to a cable car named after Emirates Airline and is dotted by a network of rentable bicycles and their docking stations emblazoned with the name of the Spanish-based bank Santander. Tourist attractions have names such as the Coca-Cola London Eye and the ArcelorMittal Orbit. Notable entertainment complexes include the O2 Arena, the Kia Oval, and, again, Emirates Stadium. The branded place names immediately signal that powerful corporations and public-private partnerships dominate the city. Soon, basic infrastructure will also carry branded sponsorships. One west London council is already seeking to sell the naming rights to a proposed new bridge over the River Thames (see Pitcher & Hurst, 2019). Corporate names are featured prominently on digital platforms like Google Maps, where the official place names of commercial entities and paid advertisers take precedence over names derived from everyday linguistic practice. The brandscape coexists with other toponymic registers. There are parks and squares named after figures from aristocratic and imperial history, and council buildings and public housing estates with names that invoke British social democracy. But increasingly, the city appears as a constellation of corporate trademarks.

This special issue of *Urban Geography* shows that, in cities as varied as Dubai, San Francisco, Timişoara, and Blenheim, the selling of naming rights is becoming ever more common. Public as well as private organizations are turning to sponsorship deals as a way to raise revenue. It is possible to purchase the names of public facilities, train stations, community centers, parks, bridges, monuments, stadiums, and other places, institutions, and infrastructure. This may seem like a harmless trend with few serious consequences. But the research in this issue demonstrates that place branding reflects the major motif of the neoliberal city: dispossession, in the form of city dwellers stripped of common identifiers, spaces, property, and institutions; governments stripped of their budgets; and cities stripped of their assets. It should therefore not be surprising that place branding is almost always contested, and often a flashpoint in struggles over urban space.

Branded corporate renaming is a specific form of toponymic commodification (Rose-Redwood, Sotoudehnia, & Tretter, this issue; also, see Light & Young, 2015; Medway & Warnaby, 2014), a concept that was developed within the broader literature on critical toponymy (see Berg & Vuolteenaho, 2009; Rose-Redwood & Alderman, 2011; Rose-Redwood, Alderman, & Azaryahu, 2010, 2018). Toponymic commodification is made possible by political-economic restructuring that occurs across multiple levels, scales, and territories. As real estate becomes more significant relative to other economic sectors, place names become increasingly useful for marketing and other purposes. But usefulness does not turn something into a commodity. To be made into property that can be traded and circulated, place names also require state action to recognize their legitimacy, regulate their sale, maintain their legal status, inscribe them on maps and official registers, and in some cases sell them directly to willing buyers. Like all instances of commodification, "toponymic capture" (Kearns & Lewis, this issue) is a collaboration between the state and capital in order to take something that had previously been unpriced and unsold and render it into something alienable and capable of generating revenue. In this respect, the selling of naming rights can be seen as a form of enclosure (see May, 2000).

Renaming a place or institution entails the creation of new interests, identities, and goals; it also means severing connections with those of the past. Sponsorship is not really about names. It is about economic relationships between companies and places or institutions, where the company provides funding in exchange for naming rights. The publicity generated by sponsorship deals might serve to launder a corporation's bad reputation or merely to gain name recognition. In any case, naming rights are the preserve of major or sole sponsors, and this establishes a unique kind of relationship. Unlike public funding (which is subject to democratic oversight), and unlike membership support models or even traditional advertising-based funding models (which draw on a combination of a large number of different sponsors), major sponsorship produces asymmetrical dependency. Sponsorships significant enough to include naming rights can lead to conditions where, for the place or organization being funded, the sponsor is too big to defy. In effect, all corporate sponsorships are privatizations to a greater or lesser extent. No matter who or what is listed as an official owner, once a corporate name is attached to a place, the corporation's and the place's interests are for practical purposes intertwined. Corporate reputations must now be respected and protected. Companies can now threaten to withdraw their sponsorship. Sponsored names are not inert labels but usable tools, which is one reason why adopting them enables so much money to flow.

One argument that runs through all of the articles in this issue is that sponsored renaming is not spontaneous or improvised. Instead, it follows established logics of contemporary urban development. To be clear, the valuation of place names is not in itself new. Some commercial names and place names have long functioned as generators of value, to say nothing of their power in political and cultural spheres. And many "heritage names" continue to be valuable today (Vuolteenaho, Wolny, & Puzey, this issue). But what is new is the scale, regularization, and routinization of toponymic commodification; the distinctive spatial projects (see Madden, 2014) and redevelopment patterns in which renaming plays a part; and the establishment of a competitive global market for sponsorship. Toponymic change can be found throughout urban history. But corporate sponsorships and the commodification of naming rights are contemporary, historically specific phenomena.

These studies demonstrate that place branding should not be seen as one monolithic project; at the same time, as a collection of practices, branded renaming is indeed deeply rooted in neoliberal urbanism. The neoliberalization of urban space is its condition of possibility. Only as a result of neoliberal austerity are municipal governments so hollowed out that the sale of naming rights seems like a fiscal lifeline. Only with neoliberal inequality are corporations so emboldened that their owners and strategists think they can override local naming traditions and lay claim to major city landmarks and social spaces. Only neoliberal entrepreneurialism has produced the relationships between the branding industry, the state, and real estate capital that make corporate renaming happen. Only under neoliberal politics do city dwellers expect that their primary mode of incorporation into the urban order is as consumers and therefore see the corporate colonization of their city as unexceptional.

It is no coincidence that these conditions are conducive to practices that coalesce specifically around entertainment and sport (Creţan, this issue; Medway, Warnaby, Gillooly, & Millington, this issue; Vuolteenaho et al., this issue). Neoliberal urbanism emphasizes consumption, experience, spectacle, and culture-led development. Mega-events and festivals are some of the prime vectors of neoliberal policy ideas (Hall, 2006; Waitt, 2008). As a result, in the past few decades, many cities have constructed new stadiums, music venues, "creative" hubs, nightlife districts, and entertainment complexes. And many of these sites have used renaming as part of their spatial strategy.

Despite its enthusiastic embrace by managers and consultants, these articles also demonstrate that sponsored place branding is no guaranteed source of easy money. The selling of naming rights carries a number of risks. There is the possibility that a corporate sponsor will fall into disrepute; as one of Rose-Redwood, Sotoudehnia, and Tretter interviewees remarks, "You don't know if the company's reputation will change, or how it will change. Look at BP, or Enron. You just don't know what that name will mean in one, five, or ten years" (this issue). There is also a risk that a renamed stadium or sports team will be seen as illegitimate and rejected by the public, as occurred with the football team Poli Timişoara (Creţan, this issue). The potential for a branded neologism to fail to catch on with paying audiences is especially significant within association football, "a game paradoxically known both for its heavyweight commercialism and as a realm of sometimes enthusiastically traditionalist fan cultures" (Vuolteenaho et al., this issue).

It is clear from this research that resistance to renaming and place branding is widespread. Movements mobilize against branded renaming for many reasons. As with other elements of neoliberalism, resistance can manifest in contradictory ways. Sometimes it appears in conservative forms, as with opposition to new names from traditionalist football supporters. Other examples of anti-renaming resistance articulate a more radical politics and build coalitions with groups organizing around, for example, housing, labor, or health. McElroy's (this issue) study of the renaming of San Francisco General Hospital after Facebook founder Mark Zuckerberg provides an example of the latter. The movement that emerged in opposition to it brought together nurses, doctors, anti-eviction activists, and others. The renaming of the hospital touched upon multiple expressions of urban inequality in the Bay Area. Not only did it reveal the ambition of the new tech elite to inscribe its names on the cityscape as thoroughly as did the robber barons of late industrial capitalism. It also directly shaped patients' interaction with doctors and nurses. One nurse, for example, reported that as a result of the renaming, patients feared for the security of their medical information (McElroy, this issue). Resisting "the Zuckerberg" became a way to protest tech-industry-fueled gentrification in San Francisco. It drew attention to the racialization of health problems stemming from displacement and housing injustice in the region. And it sought to contest the creeping role of datafication and digital surveillance in everyday life both off- and online. While protest against the renaming of the General did not succeeded in removing the Zuckererg name from the hospital, it did become a way to politicize emerging intersections between technology, philanthropy, inequality, and urban space.

McElroy's analysis of "techno-imperialism" points to some of the forms which urban toponymic struggles are likely to take in the future. As long as urban regions remain marked by public austerity and private luxury, place branding and renaming will continue to be tools used by corporations and governments. Just as sponsorship deals migrated from sports teams to landmarks to infrastructure, we can expect toponymic and semiotic commodification to expand into new realms and new scales, seeping farther into the corners of everyday urban life. As tech capitalists, real estate developers, sports team owners, and other elites seek to cement their place as the new ruling class, they will continue to appropriate, rename, and reinscribe urban space. Tech firms are likely to be central players in this process, due to their wealth, their urban-centric strategy for office location and recruitment, their predilection for branding and hype, and their leading role in shaping geographical nomenclature. But they are obviously only one part of this story. Capitalism today is strongly oriented towards the redevelopment and revalorization of urban space. This implies the dispossession and displacement of existing populations and uses. In the resulting struggles, there is every reason to think that toponymy will continue to be weaponized.

Ultimately, these conflicts reflect foundational questions about the politics of urban life: Who rules the city? Who decides how urban space will be used and controlled? Place names matter because they are tokens and tools in projects that seek to answer those questions. There is a reason that a city filled with corporate monikers became a common trope in dystopian novels and films. The ubiquity of branded, sponsored names indicates that the city is ruled by corporations, property owners, and their allies. Not all toponymic struggles are created equal, of course. Some have higher stakes than others. But as this special issue demonstrates, in various ways, they all touch upon fundamental issues regarding ownership, authority, property, and power in urban space.

Disclosure statement

No potential conflict of interest was reported by the author.

References

Berg, Lawrence, & Vuolteenaho, Jani (Eds.). (2009). *Critical toponymies: The contested politics of place naming*. Farnham: Ashgate.

Creţan, Remus. (this issue). Who owns the name? Fandom, social inequalities and the contested renaming of a football club in Timişoara, Romania. *Urban Geography*.

Hall, C. Michael. (2006). Urban entrepreneurship, corporate interests and sports mega-events: The thin policies of competitiveness within the hard outcomes of neoliberalism. *The Sociological Review*, *54*(2_suppl), 59–70.

Kearns, Robin, & Lewis, Nicolas. (this issue). City renaming as brand promotion: Exploring neoliberal projects and community resistance in New Zealand. *Urban Geography*.

Light, Duncan, & Young, Craig. (2015). Toponymy as commodity: Exploring the economic dimensions of urban place names. *International Journal of Urban and Regional Research*, *39* (3), 435–450.

Madden, David. (2014). Neighborhood as spatial project: Making the urban order on the downtown Brooklyn waterfront. *International Journal of Urban and Regional Research*, *38* (2), 471–497.

May, Christopher. (2000). *A global political economy of intellectual property rights: The new enclosures?* New York: Routledge.

McElroy, Erin. (this issue). Data, dispossession, and Facebook: Techno-imperialism and toponymy in gentrifying San Francisco. *Urban Geography*.

Medway, Dominic, & Warnaby, Gary. (2014). What in a name? Place branding and toponymic commodification. *Environment and Planning A*, *46*(1), 153–167.

Medway, Dominic, Warnaby, Gary, Gillooly, Leah, & Millington, Steve. (this issue). Scalar tensions in urban toponymic inscription: The corporate (re)naming of football stadia. *Urban Geography*.

Pitcher, Greg, & Hurst, Will. (2019, March 4). Council hopes sale of naming rights will deliver long-awaited Thames bridge. *Architects' Journal*.

Rose-Redwood, Reuben, & Alderman, Derek. (2011). Critical interventions in political toponymy. *ACME: an International E-Journal for Critical Geographies*, *10*(1), 1–6.

Rose-Redwood, Reuben, Alderman, Derek, & Azaryahu, Maoz. (2010). Geographies of toponymic inscription: New directions in critical place-name studies. *Progress in Human Geography*, *34*(4), 453–470.

Rose-Redwood, Reuben, Alderman, Derek, & Azaryahu, Maoz (Eds.). (2018). *The political life of urban streetscapes: Naming, politics, and place*. New York: Routledge.

Rose-Redwood, Reuben, Sotoudehnia, Maral, & Tretter, Eliot. (this issue). Turn your brand into a destination": Toponymic commodification and the branding of place in Dubai and Winnipeg. *Urban Geography*.

Vuolteenaho, Jani, Wolny, Matthias, & Puzey, Guy. (this issue). "This venue is brought to you by . . . ": The diffusion of sports and entertainment facility name sponsorship in urban Europe. *Urban Geography*.

Waitt, Gordon. (2008). Urban festivals: Geographies of hype, helplessness and hope. *Geography Compass*, *2*(2), 513–537.

Index

Note: Page numbers in **bold** indicate tables, and in *italics* indicate figures.